AGRIBUSINESS AS THE FUTURE OF AGRICULTURE

The Sugarcane Industry under Climate Change in the Southeast Mediterranean

AGRIBUSINESS AS THE FUTURE OF AGRICULTURE

The Sugarcane Industry under Climate Change in the Southeast Mediterranean

Youssef M. Hamada, PhD

Agricultural Economics Research Institute, Egypt

APPLE
ACADEMIC
PRESS

Apple Academic Press Inc.
4164 Lakeshore Road
Burlington ON L7L 1A4
Canada

Apple Academic Press Inc.
1265 Goldenrod Circle NE
Palm Bay, Florida 32905
USA

First issued in paperback 2021

© 2021 by Apple Academic Press, Inc.

Exclusive worldwide distribution by CRC Press, a member of Taylor & Francis Group

ISBN-13: 978-1-77188-842-4 (hbk)
ISBN-13: 978-1-77463-900-9 (pbk)
ISBN-13: 978-0-42932-170-2 (ebook)

Library and Archives Canada Cataloguing in Publication

Title: Agribusiness as the future of agriculture : the sugarcane industry under climate change in the southeast Mediterranean/ Youssef M. Hamada, PhD, Agricultural Economics Research Institute, Egypt.

Other titles: Diseases of field crops (Burlington, Ont.)

Names: Hamada, Youssef M., author.

Description: Includes bibliographical references and index.

Identifiers: Canadiana (print) 20200197673 | Canadiana (ebook) 20200197738 | ISBN 9781771888424 (hardcover) | ISBN 9780429321702 (ebook))

Subjects: LCSH: Sugarcane industry—Mediterranean Region. | LCSH: Sugarcane—Climatic factors—Mediterranean Region. | LCSH: Agricultural industries—Mediterranean Region. | LCSH: Agriculture—Economic aspects.

Classification: LCC HD9118.M44 H36 2020 | DDC 338.1/73610918224—dc23

Library of Congress Cataloging-in-Publication Data

Names: Hamada, Youssef M., author.

Title: Agribusiness as the future of agriculture : the sugarcane industry under climate change in the southeast Mediterranean / Youssef M. Hamada.

Description: Palm Bay, Florida, USA : Apple Academic Press, [2021] | Includes bibliographical references and index. | Summary: "This book, Agribusiness as the Future of Agriculture: The Sugarcane Industry under Climate Change in the Southeast Mediterranean, reviews the challenges of agribusiness in the Southeast Mediterranean sea. As climate change creates new risks to human populations and food security, a better analysis is needed to understand this new level of uncertainty and to understand how it will impact agriculture and its relationship with economies, livelihoods, and development. Africa emits low greenhouse gases but is hit the hardest by global warming, posing a serious challenge to increasing agricultural productivity in the Southeast Mediterranean sea. The author focuses on sugarcane cultivation in Egypt to illustrate the impact of climate change on agribusiness. Sugarcane in Egypt is used as a local food source, for international trade, for the balance of payments, for land and water use, and as a basic product for food and fiber manufacturing. Hence every aspect of the economic structure of Egypt relates to agriculture. The book examines the causes and effects of climate change on agribusiness in the Southeast Mediterranean region, such as its effect on prosperity and net farm incomes crop yields. It considers how to promote agribusiness development in the area and the potential to alleviate poverty in rural areas. It looks at the future of the sugarcane industry in Upper Egypt as a case study of agribusiness"-- Provided by publisher.

Identifiers: LCCN 2020010230 (print) | LCCN 2020010231 (ebook) | ISBN 9781771888424 (hardcover) | ISBN 9780429321702 (ebook)

Subjects: LCSH: Sugarcane industry--Mediterranean Region. | Sugarcane--Climatic factors--Mediterranean Region. | Agricultural industries--Mediterranean Region. | Agriculture--Economic aspects.

Classification: LCC HD9118.M44 H36 2021 (print) | LCC HD9118.M44 (ebook) | DDC 338.4/13610962--dc23

LC record available at https://lccn.loc.gov/2020010230

LC ebook record available at https://lccn.loc.gov/2020010231

Apple Academic Press also publishes its books in a variety of electronic formats. Some content that appears in print may not be available in electronic format. For information about Apple Academic Press products, visit our website at **www.appleacademicpress. com** and the CRC Press website at **www.crcpress.com**

About the Author

Youssef M. Hamada, PhD
Emeritus Professor of Agricultural Economy,
Agricultural Economics Research Institute,
Agricultural Research Center, Egypt,
E-mail: youssef_hamada@yahoo.com

Youssef M. Hamada, PhD, is an Emeritus Professor of Agricultural Economy at the Agricultural Economics Research Institute, Agricultural Research Center (A.R.E.), Egypt. In addition, he was a researcher at the Central Laboratory for Design and Statistics Analysis Research. He received his BSc in Agriculture Economics, his MSc in the economics of the bread industry at A.R.E. at Al-Azhar University, and his PhD in the possibilities of the bread-industry development at A.R.E. from the University of Al-Azhar, Egypt. He received his Diploma in Translation Arabic—English from the Translation Center, American University, Egypt. He was awarded training from ECARDA, Aleppo, Syria (training course for building databases and electronic websites on the Internet), and an International Training Workshop on Economic Development Policies and Practices for Desertification and Dry Land for Affected African Countries from the Chinese Academy of Forestry, Beijing, China. He has published more than 30 papers in local and international journals as well as two book chapters and a book.

Contents

Abbreviations

AU	African Union
BAU	business-as-usual
BB	bud burst
CAADP	Class Africa Agriculture Development Program
CCFS	climate change and food security
CCVI	climate change vulnerability indicator
CDM	clean development mechanism
CFCs	chlorofluorocarbons
CH_4	methane
CO_2	carbon dioxide
COMESA	Common Market for Eastern and Southern Africa
ECOWAS	Economic Community of West African States
ENSO	El Niño southern oscillation
EU	European Union
FAO	Food and Agriculture Organization
FDI	foreign direct investment
FIVIMS	food insecurity and vulnerability data and mapping system
FOSI	future of sugar industry
GCOS	global climate observing system
GDP	gross domestic product
GECAFS	global environmental modification and food systems
GHG	greenhouse-gas
IBRD	International Bank for Reconstruction and Development
IFC	International Finance Corporation
ILO	International Labor Organization
IOC	Intergovernmental Oceanographic Commission
IPCC	Intergovernmental Panel on Climate Change
ISIC	International Standard Industrial Classification
KP	Kyoto Protocol
LDCs	least developed countries
MDGs	millennium development goals

N_2O	nitrous oxide
NAPA	National Adaptation Programs of Action
NPP	net primary production
OECD	Organization for Economic Cooperation and Development
PAHs	polycyclic aromatic hydrocarbons
RCPs	representative concentration pathways
SADC	South African Development Community
SIDS	Small Island Developing States
SMEs	small- and medium-sized enterprises
SPC	Sahel Ian Pesticide Committee
SSA	sub-Saharan Africa
TFP	total factor productivity
UN	The United Nations
UNESCO	United Nations Educational, Scientific, and Cultural Organization
UNFCCC	United Nations Framework Convention on Climate Change
USA	United States of America
VPD	vapors pressure deficit
WCRP	World Climate Analysis Program
WFS	World Food Summit
WMO	World Meteoric Organization

Acknowledgments

I acknowledge the editor and thank the reviewers for the advice and help with this manuscript. This work is dedicated to my sister, my brothers, and in memory of my parents. Errors and omissions remain the responsibility of the author.

—Prof. Youssef M. Hamada

Preface

This book reviews the troubles of agribusiness as a future of agriculture in the Southeast Mediterranean Sea. As climate change creates new risks, a better analysis is needed to understand a new level of uncertainty. In order to plan for disasters, we need to understand how climate change will impact on economies, livelihoods, and development. The smart agricultural policy has also developed toward risk management programming that helps farmers deal with short-term income fluctuations as a result of risks largely outside their control. But the risks in agriculture today are greater and more complicated than ever before. International competition is fierce.

Africa emits low greenhouse gases but hit hardest by global warming. Thus, developed nations shouldn't solely bear responsibility for inflicting heating however additionally fight against it. To tackle heating, the Kyoto Protocol set target levels for industrialized nations to scale back gas emissions; but, the achievements created to this point, particularly with relation to a lot of just distribution of fresh Development Mechanism comes, and are not optimal. Climate change poses an extra, serious challenge to increasing agricultural productivity in the Southeast Mediterranean Sea.

Agriculture is a major economic issue in Egypt. It's an issue as a local food source, for international trade, for the balance of payments, land use, and water use and as a basic product for food and fiber manufacturing. Hence, every aspect of the economic structure of Egypt relates to agriculture.

Sugar is one of the main substrates of the human diet. Sugar business is one among the oldest industries in Egypt. The anticipated negative impacts for the Ethiopian Renaissance dam on downstream will be more pronounced for Egypt because it virtually depends on the Nile. Because of Egypt is being within the heart of the arid region and deficit of water resources additionally to the speedy increase of population which requires continued growth within the production of sugar to fill the native need, it absolutely was tough to expand the cultivation of sugarcane as a result of it needs massive quantities of water and is against the Egyptian water policy. Due to the supply of the previous conditions in southern Egypt governorates, they became the most home to the cultivation of sugarcane and also the home of sugarcane industry Egypt, significantly within the governorates of Sohag, Qena, and Aswan. There are eight sugar factories in Egypt distributed within the major governorates that manufacture sugarcane.

To achieve equity and efficiency in the sugar industry, nonlinear programming model was formulated to focus on the scientific linkage between water, equity and full capacity of sugarcane industries. Future of sugar industry and water scarcity model in Upper Egypt for maximize revenue of farms, maximize efficiency of factories, maximize efficiency of black honey factories, maximize efficiency of sugarcane juice shops, maximize efficiency of sugarcane seeds, minimize lost in farms, maximize labor wages, minimize water uses in planting, minimize energy consumption in cultivation, and minimize CO_2 emission in cultivation sugarcane crop in Upper Egypt as a case study of agribusiness.

According to the financial and economic analyses, the annual internal rate of return became higher than the existing model. The absolute risk of optimum cultivation reduced. All factories are working by 100% efficiency. The proposed model provided low greenhouse gases emission and reduced water consumption. Overall, as a result of an optimal sugar-cropping pattern, Egyptian sugar productions would increase.

CHAPTER 1

Introduction

ABSTRACT

Agribusiness could be a comprehensive concept shelters input suppliers, agro-processors, exporters, traders, and retailers. Business delivers inputs to farmers and joins them to customers through the handling, financing, storage, processing, marketing, transportation, and delivering of agro-industry product and maybe disintegrated further into four main teams (Yumkella et al., 2011):

1. Agricultural input business for growing agricultural production, like agricultural machinery, tools, and equipment; pesticides, fertilizers, insecticides; irrigation systems and connected equipment;
2. Agro-industry: Like food and beverages industry; tobacco product, animal skin and animal skin products; textile, footwear, and garment; wood and wood product; rubber products; yet as industry products supported agricultural materials;
3. Instrumentation for process agricultural raw materials, as well as machinery, tools, storage facilities, cooling technology, and spare parts;
4. Numerous services, financing, marketing, and distribution companies, as well as storage, transport, ICTs, packaging materials and style for higher promoting and distribution.

Therefore, agribusiness is a term accustomed mean farming and all the opposite industries and services that represent the availability chain from the farm through process, wholesaling, and merchandising to the buyer (from farm to fork within the case of food products) (Yumkella et al., 2011). Agro-industry includes all the post-harvest activities that are concerned within the transformation, preservation, and preparation of agricultural production for go-between or final consumption of food and non-food product (Wilkinson and Rocha, 2009). It consists of six main teams in step with the International Standard Industrial Classification (ISIC), particularly food and beverages; tobacco products; paper and wood products; textiles, footwear, and apparel; animal skin product; and rubber products. The term captures various primary and secondary post-harvest activities, starting from basic village-level trade goods preparation to fashionable industrial processes and involving wide differing levels of scale, quality, and labor, capital, and technology intensity. Food process industries tend to dominate this sector in developing countries, as well as the Southeast Mediterranean. Rao (2006) teams food process industries into three categories: primary—those that involve the fundamental process of natural turn out, as an example, cleaning, grading, and de-husking; secondary—those that embody easy or elementary modification of natural turn out, as an example, chemical process of edible oils; and tertiary—those that embody some sort of advanced modification to the natural turn out like creating it into edible product like tomatoes into ketchup, farm product into cheese, etc.

The agri-food system encompasses the interlinked set of activities that run from seed to table, as well as agricultural input production and distribution, farm-level production, raw product assembly, processing, and promoting. It encompasses the worth chains for various agricultural and food products and inputs and also the linkages among them. The agri-food system is additionally a shorthand term for agriculture and connected agro-industries. Whereas most of the analysis refers expressly to it a part of this expanded agriculture that produces food, several of the conclusions apply equally well to those elements of agriculture and agro-industry that turn out non-food products like fibers and biofuels.

Agro-processing is that the subset of industrial that processes raw materials and intermediate products derived from the agricultural sector. Agro-processing business, therefore, suggests that reworking products originating from agriculture, biological science, and fisheries (FAO, 1997).

This book reviews the trouble of agribusiness as a future of agriculture in the Southeast Mediterranean Sea. There is no problem in agriculture as it is, but there is a concern in the Southeast Mediterranean Sea about the future of agriculture in this place as it is. In this book, the author has presented the most views of the specialists in agriculture and agribusiness to explain the significance of agribusiness so that the economic growth in the region can be completed. In Chapter two, the agribusiness for the southeast Mediterranean's prosperity is presented. In the third chapter, the author has presented the views of the majority of specialists in climate change, causes and effects to clarify the significance of causes, and effects of climate change in order to complete the economic development in that region. In Chapter four, the author has presented the most views of the climate change specialists in Africa to explain the significance of climatic impacts on global warming in Africa. In Chapter five, the author has presented the most of the views of the majority of specialists in agriculture and development to clarify the significance of climatic impacts on crop yields and net farm incomes in Africa. As well as in Chapter six, the author has presented the most views of the specialists in climate change and food security (CCFS) to clarify the significance of climate change as an important impact on food security in Africa. In Chapter seven, the author has presented the most of the views of the specialists in future threats to agricultural food production to clarify the significance of agriculture to be able to complete the food production in Southeast Mediterranean Sea, and then moved to the problem of Egypt and the significance of agriculture in that state as the outlet of North Africa. After that, the book moved in Chapter eight to present the most of the views of the specialists in promoting agribusiness development in the Southeast Mediterranean Sea to clarify the significance of agribusiness to be able to complete development in the Southeast Mediterranean Sea. The book presented the opinions of the most of agribusiness specialists about this critical situation, which is very dangerous in Chapter nine, as it came without change or deletion, which ended with the confirmation of the riskiness of the agribusiness situation in Southeast Mediterranean Sea, especially to potential alleviate poverty in rural areas for climate change adaptation.

As a professor of agricultural economics specializing in providing solutions to the problems of the agricultural economy through the creation of mathematical models, in the tenth chapter, the author has provided a solution to the seriousness position of the agribusiness situation in Southeast Mediterranean Sea, especially to face future of sugarcane industry in Upper Egypt as a case study of agribusiness. This problem and how to decrease the risk of sugarcane production in the filling period of the Renaissance dam in Ethiopia and how to manage the sugarcane industry in Upper Egypt in that period with the lowest rate of risk and solution corresponds to that stage as described in the book. The outcome of the mathematical model was the potentiality of adaptation in the agricultural sector with that phase, increased sugarcane production to prevent famine in sugar, the potentiality of employing full agricultural employment in Upper Egypt, the potentiality of limiting carbon dioxide emissions, reducing pollution, and preserving the environment.

1.1   BACKGROUND

Climate change has the doable to worsen conflict, cause humanitarian crises; disrupt folks; destroy livelihoods; setback growth and also the competition conflicting to poorness for countless people across the world. It's foreseen that over 50 million folks within the Nile Delta in North Egypt can be forced to flit as their homes are affected by saltwater prevalence from rising sea levels. Entire populations of some low-level land states, like the Nile Delta or the Southeast Mediterranean might need to be resettled. In countries like Egypt, wherever over half the population depends on agriculture, climate-induced risks, like the insufficiency of water, which caused over billions in agricultural losses, can still create a potential for harm. Similarly, climate risk assessments in Africa show that changes in precipitation patterns, floods, and drought may place human health in danger by increasing the prevalence of respiratory and water-borne diseases and malnutrition disease. Similarly, climate risk assessments in Southeast Mediterranean show that changes in precipitation patterns, floods, and drought may place human health in danger by increasing the prevalence of respiratory and water-borne diseases and malnutrition. Long-run progressive changes can mean that folks all over should learn to adapt to weather or precipitation patterns dynamic or shifts in ecosystems that humans depend on for food. Maybe a lot of worrying, however, is that climate variability and alter also will bring unpredictable weather patterns which will, in turn, lead to a lot of extreme weather events. Heat waves, droughts, floods, and violent storms can be rather more common within the decades to return. Global climate change is loading the dice and creating extreme weather events a lot of possibilities. These disasters can undermine the property of development and render some practices, like sure kinds of agriculture, unsustainable; some places uninhabitable, and a few lives unlivable. As global climate change creates new risks, more robust analysis is required to grasp a brand new level of uncertainty. So as to arrange for disasters, we want to grasp however, global climate change can impact on economies, livelihoods, and development. We want to grasp, however, the possible changes in temperature, precipitation, yet because the frequency and magnitude of future extreme weather can have an effect on any sector, as well as agriculture, water use, human, and animal health and also the diverseness of wetlands (IISD, 2013). The sensible agricultural policy has conjointly developed toward risk management programming that helps farmers accommodate short-run financial gain fluctuations as a result of risks for the most part outside their management. However, the risks in agriculture these days are larger and a lot of sophisticated than ever before. International competition is fierce. Technological enhancements are increasing world production and driving down real trade goods costs. Public demand for higher food safety standards and higher environmental practices needs new investments within the food system. Advances in science and technology are raising ethical and moral questions on the method food will and will be made. At the identical time, sensible agriculture itself has ne'er been a lot of numerous, starting from specialty crops planted in little plots to grain farms covering thousands of hectares. In between being livestock operations of all sizes, greenhouses, organic farms and a growing variety of agricultural businesses is catering to distinctive client demands? Their surroundings that are stern new approaches to, however, business is conducted on the farm and consequently, how governments conduct agricultural policy (Hamada, 2016).

1.2 PURPOSE OF THE BOOK

In sub-Saharan Africa (SSA), industrial has been growing at a slower rate than the general gross domestic product (GDP) since 1990 suggesting delayed industrial take-off and premature de-industrialization in some countries. Premature de-industrialization, triggered partly as a consequence of the structural adjustment policies that tackled unsustainable industrialization ways, is all the same a worrying sign since it implies that surplus labor either remains in low-productivity agriculture or shifts to low-productivity informal producing or service activities. Consequently, producing has not served as a dynamic supply of development in SSA because it has drained alternative developing regions, primarily to weak industrial capabilities, inadequate institutions, and infrastructure constraints (Yumkella et al., 2011).

Against this background, what role will agribusiness play within the development of Africa, as Southeast Mediterranean Seacoast? The question demands an urgent answer for three main reasons. First, though the economic process since 1995 has been stronger than within the preceding 20 years, absolute poorness within the region continues to extend. In 1990, the baseline for the millennium development goals (MDGs), 295 million folks in SSA was calculable to be living on less than $1.25 a day, a number that had up to 388 million by 2005. The corresponding figure for $2 every day had up from 390 million to 555 million. Second, from this information, it's clear that not solely is growth inadequate to reverse this trend however conjointly that the pattern of growth—largely obsessed on capital-intensive, resource-depleting energy and mining expansion—has didn't turn out the specified trickle-down effects necessary to scale back the poverty headcount. Third, most Africans board rural areas and agriculture remains the only largest supply of employment and financial gain. Agriculture contributes 15% of GDP, virtually a simple fraction of total employment (64.7%), and accounts for over 75% of domestic trade by price providing a living for the majority of the economically active population. Agriculture plays a good a lot of dominant role within the lives of the poor, who still be primarily rural and either directly engaged in farming, or dependent on activities connected thereto. Luckily, there are indications that in some African countries agricultural development and productivity enhancements are quickening over the last one or two decades. Following many decades of relative neglect by governments and donors, there's an emerging consensus that a lot of determined efforts should be created to develop African agriculture through inflated intervention and investment. Each is required to accelerate economic development and poorness reduction whereas causative to achieving the broader MDGs (Yumkella et al., 2011).

As an economy develops over the long run, the share of agriculture, in each of gross domestic product and employment, declines. This method has long been apparent in SSA wherever agriculture's share of gross domestic product has fallen from 43% in 1965 to 12% in 2008 (World Bank, 1989; World Bank, 2009), implying that there's very little field for agriculture—narrowly-defined as crop and livestock production—to drive output growth and poorness reduction. However, agriculture's contribution are increased considerably by strengthening linkages with industry, through agro-processing and value-addition downstream of farms, within the provision of farm inputs upstream, and in improved post-harvest operations, storage, distribution and provision that are essential components of business value chains. This offers a route to economic development and poorness reduction, yet because the structural transformation of economies and also the improvement of technical skills and capability. Wilkinson & Rocha (2009) have shown by trial and error that the magnitude relation of gross domestic product generated by agribusiness to it generated by farming will increase from 0.57 for a sample of nine "agriculturally-based countries" (all in SSA) to 1.98 for a collection of eleven "transforming countries" (mainly Asian) and to 3.32 for twelve "urbanized countries." For the US, the ratio stands at thirteen, whereas in agricultural countries that haven't undergone a structural transformation, 63% of the value added within the agrifood system was created on the farm, in the US, farming accounted for less than 7% Input producers, agro-industry, truck companies, restaurants staff, et al. created the value added another within the United States agrifood system, implying that agribusiness is considerably necessary for value addition and economic prosperity.

1.3   SCOPE OF THE BOOK

For centuries, agriculture has driven economic development in countries across the world, and Southeast Mediterranean nations are following the identical path out of poorness. With agriculture accounting for 65% of the continent's employment and 75% of its domestic trade, it's possible to drive Africa's economic development for years to come. Smallholders are the backbone of that effort. New and evolving markets hold the promise of larger profits for smallholder farmers. Feeding the quickly growing urban population would require a lot of and better quality agricultural commodities. Urban customers also will increase demand for processed agricultural products, therefore adding value to farmers' outputs can take center stage in years to come. This can offer profitable opportunities not only for the ladies and men who grow the food, but for a large vary of rural employees, particularly the emerging generation of young's (Yumkella et al., 2011).

A key initiative in exploiting these opportunities is recognizing smallholder farms as agribusinesses, irrespective of their size or weight. Sadly, too several little agribusinesses in Southeast Mediterranean are neither productive nor profitable. There are two vital reasons why they continue to be at bay during a cycle of subsistence. First, their yields are too low to come up with marketable surpluses; as a result of they lack access to fashionable technology and productive assets. Second, farmers cannot get their produce to markets, because of the rareness of roads and linkages between farm-level production and downstream activities, like processing and promoting. African agriculture and agribusiness should be reworked to satisfy the stress of the twenty-first century, and this book outlines the crucial ingredients. I feel we want to spark an agribusiness and agro-industrial revolution for the good thing about rural areas. Such a revolution can bring sustained investment within the entire agribusiness value chain, which, in turn, can raise productivity and yields, improve competitiveness and increase profits. By implementing the thoughtful, practical ideas reported in this book, we will so use agribusiness to make prosperity for Southeast Mediterranean which suggests that prosperity for the ladies and men who feed the Southeast Mediterranean Sea's folks.

1.4 CONCLUSION

Urbanization, the exit of labor from agriculture and a decline in agriculture's contribution to GDP have traditionally characterized the structural transformation of socio-financial structures. In today's high-income countries, this technique caused the emergence of an city middle magnificence and a large shift in food preferences in the direction of meat and dairy merchandise. Although proof continues to be sparse and researches are ongoing, the same technique seems to be occurring in low- and middle-profits nations. At the equal time, demographic pressure in those countries is increasing. Together, those dynamics exchange meals structures in numerous ways, and these changes, in turn, drive further structural transformation. While population growth increases the demand for agricultural products and stimulates farming activities, urbanization requires food to be easily stored and transported. Thus, food processing has end up a key factor inside the transformation of food structures. It has delivered with it the standardization of agricultural output and, in many cases, the concentration of number one manufacturing and the consolidation of farmland. Many smallholder farmers have grown to be landless agricultural workers, or have migrated to cities and cities in search of employment, accelerating urbanization.

Agriculture and food manufacturing are increasingly supplying city and peri-city supermarkets. Value chains are steadily characterized by using the vertical coordination, and in some times the integration, of number one manufacturing, processing and distribution; the automation of large-scale processing; and better capital and understanding intensities. A complete global assessment of these transformations, particularly in the wholesale and retail segments of the cost chains, is difficult, attributable to the lack of effortlessly accessible and comparable data. However, some traits by companies of nations and regions may be inferred from current literature. Between 2001 and 2014, the proportion of processed food allotted through supermarkets notably improved in upper middle-income international locations, from less than 40–50%. In the equal period, it grew from round 72% to 75% in high-earnings international locations. In decrease center-earnings countries, it grew from 22 to 27% between 2001 and 2008, with no similarly alternate among 2008 and 2014 (Global Panel on Agriculture and Food Systems for Nutrition, 2016). A different image emerges for sparkling food. Over the last 10 years, the percentage of fresh meals distributed via supermarkets has remained under 50% in high-earnings countries, beneath 30% in upper center profits international locations and round 10% in decrease-middle earnings nations (Global Panel on Agriculture and Food Systems for Nutrition, 2016).

This chapter provides a brief overview of background of the book, purpose of the book, scope of the book, agribusiness for southeast Mediterranean's prosperity, climate change, cause and effects, climatic impacts on global warming, climatic impacts on crop yields and net farm incomes, climate change and food security: an overview, future threats to agricultural food production, promoting agribusiness development in southeast Mediterranean sea, potential alleviate poverty in rural areas for climate change adaptation, future of sugarcane industry in upper Egypt as a case study of agribusiness.

KEYWORDS

- agri-food system
- agro-industries
- biofuels
- global warming
- International Standard Industrial Classification (ISIC)
- raw materials

REFERENCES

Annan, K. A., (2009). *The Anatomy of a Silent Crisis, Global Humanitarian Forum*. Global humanitarian forum, chair, steering group, human impact report: Climate change.

EEA, (2015). *Atmospheric Greenhouse Gas Concentrations*. European environment agency, EEA, Copenhagen.

Ellis, S., (2008). *The Changing Climate for Food and Agriculture*. Institute for Agriculture and Trade Policy in Minneapolis, Minnesota. ©2008 IATP. All rights reserved.

FAO (1997). Food and Agriculture Organization of the United Nations Available at: http://www.fao.org/docrep/w5800e/w5800e00.htm.

FAO, (2008). Food and Agriculture Organization of the United Nations, Rome. *Climate Change and Food Security: A Framework Document*.

Global Panel on Agriculture and Food Systems for Nutrition (2016). Food Systems and Diets: Facing the Challenges of the 21st Century, London.

Hamada, Y. M., (2016). *Risk Management in Agriculture*. Driving agribusiness with technology innovations. www.igi-global.com (accessed on 4 January 2020).

IISD, the International Institute for Sustainable Development (2013). Climate risk management for agriculture in Peru: focus on the regions of Junín and Piura, the International Institute for Sustainable Development, United Nations Development Programmed, January 2013

International Institute for Sustainable Development (IISD), (2013). *Climate Risk Management for Agriculture in Peru: Focus on the Regions of Junín and Piura*. The International Institute for Sustainable Development, United Nations Development Program.

Rao, K. L. (2006). Agro-industrial parks experience from India. Agricultural and Food Engineering Working Document 3, Rome: FAO. Available at: http://www.fao.org/ag/ags/publications/docs/AGST_WorkingDocuments/J7714_e.pdf (accessed on 4 January 2020).

Rocket Scientist, (2007). *Climate Change and Agriculture Country Note*. The Former Yugoslav, Republic of Macedonia. www.worldbank.org/eca/climateandagriculture (accessed on 4 January 2020).

Saleemul, H., (2009). *The Anatomy of a Silent Crisis, Global Humanitarian Forum*. Chair, steering group, Human impact report: Climate change.

Stern, N., Peters, S., Bakhshi, V., Bowen, A., Cameron, C., Catovsky, S., Crane, D., et al., (2006). *Stern Review: The Economics of Climate Change*. HM treasury, Chapter 6, London. http://www.hm-treasury.gov.uk/stern_review_report.htm (accessed on 4 January 2020).

UNIDO, FAO, and IFAD, (2008). *The Importance of Agro-Industry for Socio-Economic Development and Poverty Reduction*. New York: UN Commission on Sustainable Development.

Von, B. J., (2007). *The World Food Situation: New Driving Forces and Required Actions*. Food Policy Report, Washington, D.C.: IFPRI.

Walter, F., (2009). *Global Humanitarian Forum*. Chair, Steering Group, Human impact report: Climate change, CEO/Director-General.

WFP, (2009). *Who Are the Hungry?*. World Food Program. http://www.wfp.org/hunger/who-are (accessed on 4 January 2020).

Wilkinson, J., & Rocha, R. (2009). Agro-industry trends, patterns and development impacts, In C. da Silva, D. Baker, A.W. Shepherd, C. Jenane, S. Miranda-da-Cruz (eds.), *Agro Industries for Development*, Wallingford, UK: CABI for FAO and UNIDO, pp. 46–91.

World Bank (1989). *Sub-Saharan Africa: From Crisis to Sustainable Growth: A Long-Term Perspective Study*. Publication summary, Washington D.C.: The World Bank.

World Bank, (2007). *World Development Report 2008: Agriculture for Development*. Washington, D.C.: World Bank.

World Bank (2009). *World Development Indicators 2009*, Washington, D.C.: The World Bank.

Yumkella, K. K., Kormawa, P. M., Roepstorff, T. M., & Hawkins, A. M. (2011). *Agribusiness for Africa's Prosperity*, UNIDO ID/440, Layout by Smith + Bell Design (UK), Printed in Austria, May 2011.

CHAPTER 2

Agribusiness for Southeast Mediterranean's Prosperity

ABSTRACT

Southeast Mediterranean economic development remains in the main commodity-based on exports of oil, minerals, and agricultural merchandises with slight or no dispensation involved. Therefore, on hastening viable and comprehensive development in Southeast Mediterranean there is an important would like for fostering a recent development approach supported exploiting the complete agribusiness potential of the Southeast Mediterranean. Figure 2.1 shows agribusiness as a future of agriculture, drying tomato in the sun in Upper Egypt. This could specialize in increasing agro-industrial value another and use on the complete agribusiness value chain in agriculture, trade, and services. This chapter analyses the challenges and possibilities of Southeast Mediterranean agribusiness at intervals this amount of dramatic changes in world agro-manufacturing marketplaces, and shapes a durable circumstance for business development as a path to Southeast Mediterranean's prosperity. This chapter displays the matter of agribusiness for Southeast Mediterranean's Prosperity as a result of the views of the foremost of specialists in agribusiness for Southeast Mediterranean's prosperity.

2.1 BACKGROUND

Agribusiness plays a vital role in economic development. In countries with low per capita gain levels, the agricultural sector usually accounts for quite half of GDP and 60 to 80% of total employment. Agribusiness could also be a priority for the UN agency (International Finance Corporation (IFC)). IFC could be a member of the World Bank cluster that the

FIGURE 2.1   Agribusiness as a future of agriculture, tomato in sunrise, Upper Egypt, (Hamada, 2018).

most important world development institution focused exclusively on the personal sector in developing countries. Established in 1956, IFC is owned by 184 member countries, a gaggle that collectively determines our policies. Operational with private enterprises in extra than 100 countries, it uses its capital, expertise, and influence to help eliminate extreme poverty and promote shared prosperity. IFC leverages the ability of the private sector to make jobs and tackle the world's most pressing development challenges. IFC's vision is that people should have the prospect to escape poverty and improve their lives, as a result of its strong role in poverty reduction and potential for broad development impact. IFC combines investments and consolatory services to help the sector address higher demand and escalating food costs in an environmentally sustainable and socially inclusive way. IFC supports initiatives for the sustainable production of agricultural commodities. The key objective of IFC's agribusiness in Southeast Mediterranean is to:

• Promote food security;
• Enhance economic development and inclusion within the sector;

- Build environmental and social property a business driver;
- Create jobs at the farm and non-farm levels.

IFC builds the skills of farmers and links them with required finance, increasing their access to weather-based insurance to guard against losses because of environmental factors. It helps farmers improve productivity, provides climate-resilient seeds, and ensures a secure area for storage of grains and agricultural manufacture through trendy storage that is engineered and operated by private sector clients. IFC collectively helps firms become lots of resources economical and reduce water footprint. IFC builds offer chain linkages, and infrastructure and facilitates market development of native supply by helping farms meet quality and quantity requirements (IFC, 2014).

2.2 SCOPE OF THE CHAPTER

Alongside its role in stimulating economic growth, agribusiness, and agro-industrial has the potential to contribute significantly to poverty reduction and improved social outcomes and an accord is rising that agro-industries area unit a decisive a part of socially-inclusive, competitive development strategies (Wilkinson and Rocha, 2008). Proof of the link between growth and poverty reduction varies to keep with the country. Spectacular economic and industrial growth in China raised 475 million people out of poverty between 1990 and 2005, though big pockets of poverty still exist in growth-oriented areas and rural communities due to structural rigidities. In sub-Saharan Africa (SSA), despite sturdy growth in recent years, the amount of people living on less than $1.25 on a daily basis inflated by 93 million throughout constant period (Montalvo and Ravallion, 2010; World Bank, 2009). The accomplishment of the MDGs in SSA has been unnatural by two factors: first, most countries haven't met the specified GDP rate to succeed in the MDG target. Secondly, labor absorption and employment intensity are low due to a degree of growth in some capital-intensive extractive sectors.

Agribusiness directly contributes to the accomplishment of three key MDG's, specifically reducing poverty and hunger (MDG1), empowering women (MDG3), and developing world partnerships for development (MDG8). Durable synergies exist between agribusiness, agricultural performance, and poverty reduction for Africa (World Bank, 2007); economical agribusiness may stimulate agricultural growth and strong linkages between agribusiness and smallholders can cut back rural poverty. Attention on worth addition in agribusiness is, therefore, central to existing strategies for economic diversification, structural transformation, and technological upgrading of African economies. Such attention can initiate faster progress towards prosperity, by poignant the bulk of the continent's economic activities and by harnessing crucial linkages between the key economic sectors. This 'people-oriented' strategy will improve the welfare and living standards of the overwhelming majority of Africans, every as producers and customers, and from the angle of employment, financial gain and food security. On the demand side, food expenditure usually represents the largest single item of home expenditure, rising to over half total expenditure for poor households in some countries, and then the potency of post-harvest operations could also be a significant determinant of the costs purchased the urban and rural poor for food, and then a really vital consider home food security (Jaffee et al., 2003).

Agro-industrial development can contribute to improved health and food security for the poor by increasing the final accessibility, choice, and process worth of food merchandise, and sectioning food to be kept as a reserve against times of shortage, guaranteeing that spare food, that essential nutrients are live consumed throughout the year. On the provision side, agro-industrial development contains a right away impact on the livelihoods of the poor each through inflated employment in agro-industrial activities, and through exaggerated demand for primary agricultural manufacture. Though varied significantly by subsector and region, agro-industry, considerably in its initial stages of development, is relatively labor-intensive, providing a range of opportunities for self and wage employment. Agro-industrial activity in Africa is in addition of times distinguished by a high share of feminine employment, ranging from 50% to as high as 90% (Wilkinson and Rocha, 2008). For example, the 'non-traditional export sector' (vegetables, fruit, and fish products), that's presently the foremost dynamic in terms of exports from SSA. Similarly, the small-scale food method and line operations present throughout lots of the continent are live usually operated predominantly by girls; a study of small-scale urban agro-processing and line enterprises in Cameroon found that over 80% were managed by women, with men being present almost exclusively in the mechanical grinding and meat preparation activities (Ferré et al., 1999). Indeed, Charmes notes that the gender bias apparent in many agro-processing activities may contribute to the underestimate of every agro-industrial activity and feminine employment in national accounting, noting that a very high share of these activities are measure undertaken as secondary activities and are measure typically hidden behind subsistence agriculture (Charmes, 2000). Aboard job creation, agro-industrial enterprises usually offer crucial inputs and services to the farm sector for those with no access to such inputs, inflicting productivity and merchandise quality enhancements and stimulating market elicited innovation through chains and networks, facilitating linkages and allowing domestic and export markets to become more reciprocally supportive (FAO, 2007).

Agro-industry is in addition amongst the foremost accessible of commercial activities of times undertaken at small-scale, with the low initial price and technological barriers to entry. SMEs keep key actors at intervals the for the foremost half informal commerce and method networks, that dominate food procurable in a lot of (newly) urban Africa, and have verified remarkably accommodative and resilient at intervals the face of a range of economic, institutional, and infrastructural challenges (Muchnik, 2003; Sautier et al., 2006).

2.3 FUTURE AGRICULTURE IN WATER SCARCITY

Water is vital to grow food; for home, water uses yet as drinking, cooking, and sanitation; as an important input into the industry; for commerce and cultural purposes; and in sustaining the earth's ecosystems. But this essential resource is beneath threat. Increasing scarcities in a lot of the world produce challenges for national and sub-national governments and for individual water users. The challenges of growing water insufficiency are measure combined by increasing costs of developing new water; degradation of soils in irrigated areas; over pumping and depletion of groundwater; pollution and degradation of water-related ecosystems; and wasteful use of already developed provides, galvanized by subsidies and distorted incentives that influence water use (Rosegrant, 2015). Growing water scarceness and water quality constraints are live a significant challenge to future outcomes in food security, notably since agriculture is anticipated to remain the foremost vital user of fresh resources altogether regions of the world for the foreseeable future, despite quick-growing industrial and domestic demand. As non-agricultural demand for water can increase, water is going to be a lot of and a lot of transferred from irrigation to various uses in many regions. In addition, the liableness of the agricultural installation will decline whereas not vital enhancements in water management policies and investments. The augmentative figure competition and water scarceness problems, at the facet of declining liableness of agricultural installation, will place pressure on food provide and still generate concerns for international food security. Ringler et al. (2016) project future water stress, showing that in 2010, 36% of the worldwide population—some 2.4 billion folks—board water-scarce regions and 22% of the world's gross domestic product (GDP) (US$9.4 trillion at 2000) is formed in water-short areas. Moreover, 39% of worldwide grain production is in water-stressed regions. Business-as-usual (BAU) levels of water productivity underneath a medium process scenario will not be spare to cut back risks and guarantee property. Under BAU, 52% of the worldwide population (4.8 billion people), 49% of worldwide grain production, and 49% (US$63 trillion) of total value are going to be in peril as a results of water stress by 2050, which may apparently impact investment selections, increase operation prices and have a sway on the competitiveness of certain regions (Ringler et al., 2016).

This chapter summarizes projections for enhancing agricultural productivity outcomes for food security, showing that, beneath the quality agriculture scenario, increasing water scarceness and various factors are live projected to steer to retardation agricultural growth and rising food costs. Then the proof is provided on the impact of water scarceness on the economic process, and relationship between climate change and water resources is summarized and an alternate scenario is examined to look at whether or not plausible can increase in water and crop productivity will give significantly higher outcomes for water and food security through agribusiness.

2.4 WATER AND FOOD SECURITY

With declining accessibility of water and restricted land which will be fruitfully brought below cultivation, growth in area can contribute little or no to future production growth. Slow growth in investment in agricultural analysis, irrigation, and rural infrastructure in developing countries unit potential to dampen productivity growth, and process can reduce the speed of growth in productivity. These supply factors, and growing population (mainly in Africa and South Asia) and rising gain, unit projected to steer to rising food prices and slow enhancements in food security. International costs of cereals unit projected to extend by 20% even whereas not temperature changes. With climate change, across a variety of general circulation models, the mean increase between 2010 and 2050 is projected to be close to near 50%. Meat prices unit projected to extend by 20% nevertheless, with a bit decline in prices once 2040 as developed countries (Rosegrant, 2016).

2.5 WATER AND ECONOMIC GROWTH

In addition to the impacts on agriculture and food security, water deficiency and water-related investments will increase economic productivity and growth. Sadoff et al. (2015) summarize many the proof on this relationship, they conclude that the affiliation between water security and method of development is intuitively clear, however, that the empirical proof of this relationship is scarce. Newer economic science analyses have mirrored between variability in precipitation and therefore the mean levels. Brown et al. (2011), cited in Sadoff et al. (2015), show that rain variability, floods, and droughts have a statistically vital negative harmful impact on all fully totally different measures of the method in SSA. Brown et al. (2013) perceive that anomalously low or high precipitation contains a negative economic result, thereby providing proof that variability in precipitation will hinder growth.

2.6 GLOBAL CLIMATE CHANGE IMPACTS ON WATER

Climate change is projected to cause substantial changes in mean annual stream flows, the seasonal distributions of flows, melting of snowpack, and the enlarged likelihood of most high or low-flow conditions. Climate change impacts on water resources embrace changes within the temporal order of water accessibility as a results of changes in glaciers, snow as in rainfall; changes in water demands as a results of enlarged temperatures; changes in surface water accessibility associate degreed groundwater storage; an enlarged choice and intensity of most standing events (droughts and floods); changes in water quality; and low-lying rise (Rose Grant et al., 2014). World Bank (2010) shows that the majority of places will experience many variable precipitations, typically with longer dry periods within; Burke et al. (2006) explained the implications on the act and natural systems area unit widespread.

The ultimate outcome of worldwide global climate change and its effects on water accessibility aren't proverbial with accuracy. Unknowns embrace in Africa locations, direction of modification (less/more precipitation), degree of modification in precipitation (low/high), modification in precipitation intensity (low/high), and temporal order (within following five years or over multiple decades). Shifting precipitation patterns and warming temperatures may increase water inadequacy in some regions whereas fully totally different areas might expertise enlarged soil-moisture accessibility, which may increase opportunities for agricultural production (Malcolm et al., 2012). However, as a result of the International Bank for Reconstruction and Development (IBRD) (2010) noted, these unsure changes unit certain to produce additional sturdy to manage the world's water, with folks feeling several of the implications of worldwide global climate change through water. International climate change will produce versatile water allocation many vital, to manage to changes within the temporal order of water accessibility, changes in water demands, changes in surface water accessibility, and extreme events.

2.7   ENHANCING AGRICULTURAL PRODUCTIVITY

Agriculture is the world's major employer. In step with the Food and Agriculture Organization (FAO), agriculture employs regarding 60% of the poor and rural personnel in the Southeast Mediterranean, of that 35% are women. Over 80% of the world's little and marginal farmers belong to this region, in recent years. Agricultural growth in the Southeast Mediterranean has been 3%, such a lot below growth rates of different economic sectors. Low productivity, inefficient markets, lack of finance, and unpredictable weather conditions hold farmers back, keeping their diligence from leading to important can increase in gain. This chapter examines the supply-side challenges of agribusiness development in the Southeast Mediterranean Sea; link the sources of growth through worth addition, yet as potential the prime movers for transforming the agro-food system.

2.8   THE CHALLENGE

Some challenges that the sector faces are:

1. **Post-Harvest Losses:** The Food and Agricultural Organization ranks Southeast Mediterranean because of the little malnourished and food-insecure region in the world. Despite being an outsized producer, most food grains go waste owing to poor storage and transport conditions.
2. **Low Productivity:** Southeast Mediterranean is a region wherever per hectare productivity of agriculture is way less than in alternative regions. This interprets into lower incomes for farmers and disproportionate use of productive land. Owing to the restricted analysis and development capability of the final public and personal sectors in developing climate-resilient and high-yielding seed varieties, farmers depend upon ancient varieties that negatively impact agricultural productivity.
3. **Lack of Quality Standards:** In several Southeast Mediterranean, countries overuse of fertilizers and over-extraction of groundwater are major issues. Further, the absence and poor implementation of internationally recognized, standards, and certification systems end in high rejection rates in international markets.
4. **Vulnerability to Climate Change:** Consistent with Climate change Vulnerability. Some countries in the Southeast Mediterranean face extreme risks from the impacts of temperature change.

2.9   THE PRIME MOVERS IN TRANSFORMING THE AGRO-FOOD SYSTEM

The competitiveness of a price chain among the agro-food system depends on the qualification of the physical transformations that occur at each individual stage of the chain (e.g., the transformation of fruit into juice) and conjointly the coordination among the various stages among that chain. Coordination failures (e.g., the failure to deliver key inputs on time) can undermine the productivity gains achieved by improved technology. This implies that each event of technologies and coordination arrangements among the vertical chain are reciprocally dependent (Boughton et al., 1995). A lot of and a lot of, competition in agro-food systems are printed not at the individual trade level (e.g., grain milling) but between completely totally different vertical chains (Boehlje and Schrader, 1998) thus, this section focuses on the five prime movers of productivity growth and competitiveness throughout the vertical chain in SSA. These are (a) the institutional and infrastructural facultative environment; (b) access to key technologies; (c) arrangements of horizontal and vertical coordination among actors at intervals the agro-food system; (d) access to markets; (e) group action capacity; and (f) access to finance (Kandeh et al., 2011).

2.9.1   THE INSTITUTIONAL AND INFRASTRUCTURAL ENABLING ENVIRONMENT

Empower surroundings for the agro-food system involves two broad elements: (a) the policy and restrictive surroundings beneath that the system operates; and (b) the supply of basic infrastructure along with power, communication, water, and transportation. The policy and restrictive environment: Africa has stride long strides over the past 20 years in up the economics environment for agro-food development (World Bank, 2007). Commercial enterprise and financial reforms have reduced inflation in most countries, raising the planning surroundings for companies. Extra competitive exchange rates have created native production and exports, notably to regional markets at intervals continent, far more competitive with the agro-food product from the outside continent and reduced interchange shortages that previously affected imports of key machinery, spare components and inputs like fertilizer. Although enterprise size is powerfully influenced by scale economy problems, interest rates play an important role to the extent that the choice among varied technologies is driven collectively by relative issue costs. Subsidizing capital for the acquisition of tractors or freshly technologies may induce firms to adopt capital-intensive, labor-displacing where labor is abundant and inexpensive whereas capital relatively valuable. Moreover, wherever property action is weak, smallholder farmers become homeless and their land condemned for large mechanized farming operations. Brazil's development of its Cerrito region, wherever as spectacular, had adverse consequences for the endemic population that can't be unmarked. Therefore, recent moves to encourage extra large-scale farming in some African countries should painstakingly ponder these risks in following the identical path (World Bank, 2009; Braun and Denizen-Dick, 2009; Grain, 2008).

Trade policy issues are hotly debated every interval in the context of protection against extra-African imports and laws with regard to intra-regional trade. Beneath discussion is whether or not African countries and economic communities should use tariffs to defend the native production of 'sensitive' products, and if so, to what degree. The Economic Community of West African States (ECOWAS) common agricultural policy involves reducing food dependency in a (very) very perspective of food sovereignty (ECOWAS, 2009) that has meant protecting West African production against imports through the imposition of a typical external tariff. Proponents of this policy argue that it will be very robust for businesses to prosper and modernize once markets are inundated by cheap (sometimes subsidized) products from abroad. This is often the acquainted infant-industry argument, of that several of the criticisms—particularly concerning the child ne'er growing up—are well-known (Christy et al., 2009). However, whereas the treaties creating African regional economic communities call for the free movement of merchandise between the communities, the reality is that there are very big non-tariff barriers—in terms of road checkpoints, expensive body procedures, and unlawful payments—that constrain regional trade and typically even intra-country trade at intervals individual member countries. Until these barriers are reduced, exaggerated defensive against extra-regional imports will simply shield the revenues of the rent-seekers, United Nations (UN) agency manufacture such barriers to regional and intra-country trade, instead of profit the entrepreneurs at intervals the agro-food system, UN agency try to grow their businesses. On the far side national mercantilism policy, there are forms of institutional, legal, and administrative factors that influence the advantage of doing business (Christy et al., 2009). These embody the advantage of contract group action and dispute judgment, protection of property rights, the costs of making a firm (becoming formal), the openness of the state to public-private partnerships to ease finance, and support of applied agricultural and agribusiness-related analysis, as fastidiously documented at intervals the World Bank's annual Doing Business reports.

Infrastructure: the convenience of key infrastructure, like roads, water, telecommunications, and electricity, is crucial to developing competitive agro-industries. The dearth of reliable and cheap power severely constrains the event of cold chains that are essential for maintaining the quality of probably high-value perishable products like fruit and farm products and interruptions of power greatly increase costs of agro-processing. Processors face the costly selection of putting the product at the process line anytime; electricity moves the plant or investment in costly generators to assure endless offer of power. Similarly, the supply, quality, and worth of water are associate more and more important take into account things and profitability of farming and agro-industry in the twenty-first century. Agro-industries and farming are typically serious users of water. Climate change, increasing population pressure, and rising energy prices (which have a sway on pumping costs) are all making water progressively valuable worldwide. However, changes in the price of water across fully totally different regions will in all probability have a sway on where big international agribusinesses like better to offer their product, giving water-rich areas in Africa a potential advantage if they will create the other elements of the enabling environment to draw in such investment. Current price structures for water typically encourage its overuse, undermining long property of water-intensive farming and agro-processing. Form of recent water-saving technologies and institutional arrangements are presently or probably out there to house these challenges. Since many of the huge aquifers and watercourse basins in SSA are shared across countries, however, developing appropriate approaches will frequently would like regional cooperation and joint management (Diao et al., 2007).

2.9.2   ACCESS TO KEY TECHNOLOGIES

Improving the competence of the transformation that takes place at each stage of the agro-food system and responding to rapidly dynamic consumer demands for varied attributes in their food (increased assurance of food safety, 'greenness,' biological process quality, etc.) wants access to improved technologies. Dennis et al. (2009) provide a shut discussion of forces driving technological development within the agro-food system and of the promising technologies on the horizon to retort to dynamic shopper demands and environmental conditions facing agro-food producers. Two key technological issues that will be crucial in serving to verify the competitiveness of the agrifood system in SSA at intervals the approaching decade stand out: (a) access to essential inputs; and (b) access to technologies—packaging, goodness control, and communication—that influence the quality of processed product and communicate that quality to customers. With relation to improved access to crucial agricultural inputs, prime priority should be accorded to measures designed to reinforce access to the crucial agricultural inputs—fertilizer, crop protection, valid seeds—that drive the quality and sustainability of the provision of agricultural raw materials to agroindustry. The flexibleness of SSA to sustainably turnover agricultural products, along with raw materials for agro-processing, is vulnerable by declining soil fertility throughout the ground. Population pressure has rendered classic techniques of soil fertility management through the use of recent techniques, as farmers are forced a lot to remain their land in continuous cultivation. In many areas, the competitiveness of farm-level production depends on mining nutrients from the soil—clearly AN unsustainable strategy (World Bank, 2009).

While increasing organic matter at the soil is one very important a component of any answer, sustained can increase in labor productivity from the low levels will exclusively be potential with really large can increase at intervals the employment of inorganic fertilizers. Still, the incentives to use fertilizer are muffled by high fertilizer costs relative to the worth of outputs. The relation of grain prices to fertilizer faced by African farmers was historically rather additional unfavorable to the adoption of fertilizer responsive fashionable varieties than was the case in Asia throughout the green revolution (Heisey and Mwangi, 1997), due partially to the dearth of economies of scale in commerce fertilizer in countries where demand is proscribed. Different factors making for expensive fertilizer provide embody high midland transport costs, that reduce farm-gate grain prices and increase farm-gate fertilizer prices, policy uncertainty regarding whether or not government will itself sell backed fertilizer, thereby discouraging non-public investment in fertilizer distribution systems, and high costs of native production as a results of issues with scale and high energy costs. In 2006, the value of fertilizer at intervals in the United Republic of Tanzania was 49% on high of that in Thailand, whereas in the African nation it absolutely was 80% higher. The variations were due largely to higher per-unit costs of transport, taxes, funding, and marketing margins (Amit, 2009).

Moreover, as a result of this little African farmland is irrigated and markets are usually dissipate, every yields and costs are lots of variable than in an exceedingly lot of Asia, increasing the money risks to farmers of mistreatment plant food. To boot, the dearth of technical information offered to most farmers regarding that fertilizer formulation is most acceptable for his or her soil conditions and crop different greatly reduces the efficiency of fertilizer use. Hence, developing lots of cost-effective fertilizer worth chains involving every of domestic production and imports is one in all the crucial challenges for building competitive agro-industries in Africa. As a result of scale, economies are so very important in every production and international trade of fertilizer, sub-regional cooperation among African countries is needed to capture lower per-unit costs of foreign and region-ally factory-made fertilizers. Morris et al. (2007) argue that reducing intra-regional protectionism, adopting common quality standards, and harmonizing approval processes to increase the dimensions of national and regional markets are crucial steps needed to allow fertilizer importers and manufacturers to capture these economies. Similar issues with sub-regional cooperation and specialization arise at intervals the manufacture of sure sorts of agricultural and agro-processing instrumentation, like trac-tors, different vital agricultural instrumentation, and specialized method instrumentation, as few SSA national markets are great enough, to support such industries at economical levels.

The ability to access and use crop protectants like pesticides safely is critically very important for the health of farmhands and customers. It's to boot essential if African agro-food producers try to access additional and additional strict markets in high-income countries, that certification and management systems guarantee such product area unit used safely, that the following output is free of harmful residues. A peculiarity of these inputs is that the high level of technical information required to firmly use them. The scope for misuse by poorly educated farmers and farm laborers is not good, particularly at the presence of weak restrictive frameworks and group action mechanisms which can cause product being foreign that are out of date or not permissible in different regions of the world, and their use in ways in which apart from those that they were designed (applica-tion on the wrong crops, at intervals the incorrect doses and/or with the wrong timing). The matter has been combined by the excess of national laws across the continent and so the limited size of individual markets that reduces incentives for foreign producers of these product to tailor the inputs to native desires or to take a position at intervals the extension of technical data on use of the merchandise to input dealers and farmers. A vital wish, therefore, is to develop lots of uniform and consistent restrictive frameworks for the importation, distribution, and use of that product. In

West Africa, the Permanent Interstate Committee for Combating Drought at intervals the Sahel (CILSS), an interstate organization that covers nine countries, has developed a system of regional review and certification of pesticides utilized in its member states that tries to handle this drawback of harmonization. The Sahel Ian Pesticide Committee (SPC), which pools scientific expertise from across its member states, reviews the technical applications of chemical manufacturers and importers obtainable of their product for specific uses and either authorize, limits, or bans the utilization of the product within the sub-region. As of January 2010, 471 totally different pesticides were fogbound in the SPC information (Sahel Institute, 2010) of that 60 were prohibited for all uses in CILSS countries and plenty of others had varied sorts of restrictions on their use. Whereas the CILSS approach illustrates variety of the potential of such regional approaches to regulation therefore on win economies of scale, the weakness has been the restricted national funding of national committees charged with group action of the laws for chemical management at intervals the individual member states (Toe, 2009; ECOWAS, 2005; Me-Nsope et al., 2008).

Improving convenience and access to raised plasma for every animal and plant production is crucial to assuring a reliable provider of quality raw materials for agro-processing. Quick dynamical shopper demands and environmental conditions (climate change, unfold of animal diseases like avian flu and so the emergence of latest crop pests) would need a current stream of latest plasma to stay up and enhance productivity. In an exceedingly countries in SSA, public support for agricultural analysis and training of scientific personnel to undertake it stagnated or withered from the Nineteen through to 2005 (Nienke and Stads, 2006). Private analysis targeted on many lucrative export crops but there are few private-public partnerships like those who have characterized dynamic agricultural analysis systems, like that of Brazil (Pardey et al., 2006). Advances in biotechnology provide the potential to retort to variety of the challenges of adaptation, productivity image improvement and development of latest product for rising new demands (e.g., for nutraceuticals), but issues regarding its safety, shopper acceptance (particularly if product are exported to high-income countries) and environmental impacts and shortages of expertise have led to solely some countries developing the safety standards necessary for the introduction and testing of bioengineered crops or animals. As of 2006, just one country in SSA, South Africa, was growing any genetically modified crops commercially (Eicher et al.,

2006), so testing of Bt21 cotton has recently begun in AN extremely few others (e.g., Burkina Faso and Mali).

The challenge for public policy with relevancy access to improved plasma revolves around three problems: providing adequate funding for national and regional agricultural analysis systems and systems of agricultural instruction (to train the subsequent generation of agricultural scientists and technicians); creating frameworks conducive to larger public-private partnerships in analysis (including breakdown knotty issues regarding possession of any holding that results from such partnerships); and developing the restrictive framework to manage use of various sorts of biotechnology in agriculture. Given the little size of most national agricultural analysis systems in the continent, over half that had fewer than 100 scientists in 2000 (Nienke and Stads, 2006), regional collaboration ought to be AN outsized a part of this effort. The revived interest in African agriculture by African governments, development partners (including the 'new philanthropists' just like the Bill and Melinda Gates Foundation) and foreign investors since 2005 offers some hope that these constraints will begin to be addressed. The question of the acceptable scale of agricultural and agro-industrial instrumentation adopted in varied operations is difficult and may need major impacts on but powerfully the agro-food system acts as a driver of job creation, as a result of the convenience and price of machinery determines, in part, the flexibleness of the system to retort efficiently to dynamic shopper demands. Unluckily, too sometimes in Africa, government leaders equate large-scale with new. The essential physics and biology of sure processes (land leveling, conversion of nitrogen into urea, tomato paste process, and first process of sugarcane) dictate that they're applied practice large-scale, capital-intensive instrumentation to get the littlest quantity cost. Insistence that such operations adopt smaller-scale, wearying methods of production would nonexclusively build them less competitive against international competitors, but come up production costs for buying firms that use their outputs as inputs into resultant stages of production (e.g., farmers exploitation urea fertilizer), thereby compromising output and employment growth in downstream segments of the agro-food system.

In things where scale economies are decisive to the manufacture of agricultural machinery and agro-processing instrumentation, regional cooperation and specialization within the location of plants, whereas fascinating, is typically tough, to realize as a result of every country wishes to attract manufacturing investment. Prospects are in all probability to be most promising at intervals the few African economies where the engineering trade is well developed, like in Egypt and South African, though competition with foreign machinery and instrumentation is extraordinarily strong. In different cases, however, a way wider vary of technological decisions may exist, with the economically best different betting on relative issue prices and so the required timeliness of operations required. One example is land preparation, where animal-traction instrumentation, hand tractors, and large-scale tractors are all decisions. The choice depends on such aspects as a result of the heaviness of the soil to be plowed, the quickness thereupon the operation should happen (for example, therefore on accommodate multiple cropping within one year), the provision of maintenance services and spare components, and so the relative costs of labor and capital. For agricultural machinery and agro-processing instrumentation, a range of simpler, lots of wearying, but economically effective, technologies are typically offered. The widespread importation into SSA of straight forward grain mills, pumps, and different agricultural technologies from India shows that after faced with unsubsidized prices, African farmers and processors sometimes decide such technologies. Shifts in shopper demand will even have an impression on the choice of scale of machinery at intervals the light-weight of multiple decisions. An example is rice grind wherever, beneath conditions of low-incomes (which not exclusively reduce labor costs but to boot limit the effective demand for high-quality rice), small-scale mills may even be rather additional economically efficient than large-scale industrial mills. Such was the case in the Republic of Mali following the liberalization of the rice milling industry in 1992, once the new very little village-based plate mills drove former state-owned industrial rice mills out of business within 2 years at intervals the workplace du Niger (Diarra et al., 2000). The little mills, however, have drawback manufacturing a uniform quality of polished rice, therefore, as incomes of Mali's very little social class has grown up, some millers are shifting to intermediate-scale roller mills which can prove lots of consistent quality rice, while not subbing giant amounts of high-priced capital for affordable native labor (Lenaghan, 2009). Critical technologies for agro-processing three different very important determinants of the aggressiveness of African agro-industries are: (a) access to and worth of acceptable packaging technologies; (b) capacities to verify and certify quality; and (c) the flexibleness to publicize their product to potential purchasers.

Although exciting new technical developments are on the horizon in packaging for the agro-food product (Dennis et al., 2009), for many very little and medium-sized agro process firms in Africa, restricted access to, and so the worth of, acceptable packaging materials seriously compromise they're competitive. Faced with these constraints, small-scale enterprises recycle packaging materials (e.g., placing their fruit preserves in previous mayonnaise jars), that present serious potential hygiene problems and precludes these products from close to the lowest-income markets where effective demand for quality is low. Inappropriate packaging reduces the quality and amount of processed foodstuff and would possibly lead to product contamination. As an example, the dearth of 'breathable' pouches for regionally created potato chips ends up in their deterioration as a result of condition, and hand-held thermal sealers of plastic sachets to boot do not consistently provide a good seal resulting in contamination (Ilboudo and Kambou, 2009).

Constraints exist at three levels: inadequate technical data among a variety of those agro-processors; the non-availability of acceptable packaging materials coupled to the requirement for large minimum order size from foreign manufacturers; and so the high price of packaging materials once offered. The second issue of control is crucial, as a result of the flexibility to make sure to customers the quality of food products is crucial to contend in international markets and more and more thus in regional and national markets in the continent equally. Still structure capability to implement traceability and food safety (e.g., Hazard Analysis and important management Points (HACCP) practices, and lack of access to laboratories and different facilities to certify that product meets quality standards, cause serious challenges to very little and medium-sized agro-industrial firms. Frequently, developing such quality-control systems represents outsized fastened costs, which suggests that it's more durable for SMEs than for large firms to pay off such investments. To handle this disadvantage, group action capacities ought to be increased aboard improved vertical and horizontal coordination among firms within the worth chains. Invariably, laboratories and certification firms ought to even be impressed.

Regarding communication, even where SMEs turn out a quality product, communicate this product to customers could also be a troublesome situation. Advertising on radio, TV, and billboards involves very important scale economies, whereas customers of most agro-processing SMEs in SSA do not usually use the web, a technology that reduces those costs. Its inability to plug their product puts SMEs at a big disadvantage. Nagai (2008) found that nearly all mothers of babies he interviewed in Accra capital of Ghana in 2007 had not detected of the infant weaning food mix, that has been created by SMEs for over 10 years using a formula developed by the Ghanaians Ministry of Health and so the world organization Children's Fund (UNICEF). In contrast, most of the mothers interviewed knew of and most popular Cereal, created by Nestlé, that's wide promoted in Ghana and sells for three times the price of Wean combine. Joint advertising of generic product through SME associations might facilitate overcome this disadvantage considerably once coupled to the event, and acceptance, of recognized symbols for certified quality, sponsored by the associations.

2.9.3 ARRANGEMENTS FOR HORIZONTAL AND VERTICAL COORDINATION

Horizontal coordination: to strengthen gain, agro-industrial firms must be compelled to figure along to match the size enjoyed by large businesses. Such horizontal coordination is commonly achieved through skilled associations of millers, food processors or farmers through the provision of services to members coaching, research, promoting facilitate and political action—to harness market power through bulk shopping for of inputs or promoting economies at intervals the sale of output. At the farm level, horizontal coordination takes the form of collective actions by cooperatives and at intervals the case of agro-industries, by skilled associations (Kandeh et al., 2011). Such organizations may additionally attempt to set costs for the product and services created by their members, though sometimes such tries square measure unsuccessful once varied members build policing difficult. Sometimes these organizations are accustomed to facilitate vertical coordination with larger firms either upstream or downstream through an exploit. Among medium and large-scale firms, associations have proved instrumental in providing joint services to their members like the cell-phone based totally agricultural brokering service offered by the Republic of Zambia National Farmers Union an association of medium and large-scale farmers. Sometimes, however, their actions reduce shopper welfare as in cases where large-scale maize millers in southern and eastern

Africa have used their lobbying influence to limit competition thereby increasing retail prices (Jayne and Jones, 1997; Christy et al., 2009).

Vertical coordination: It involves creating incentives for firms on the price chain to integrate their operations in an extremely reciprocally advantageous fashion. This may take the form of open markets, exploit arrangements, grading standards, strategic alliances, worth chain partici-pant councils and shared possession. Vertical coordination acknowledges that the gain of activities at one level of a price chain (milling) depends critically on decisions taken at different levels, just like the choice of a variety of grain. Consequently, for the worth chain to be profitable for all, participants' actions at altogether different stages ought to be harmonious. If all the conditions of the economist's model of wonderful competition could also be met, relative prices alone would assure such coordination, as they could synthesize all the data out there throughout the system concerning shopper preferences and additionally the marginal costs of responding to those preferences. At intervals the world, however, many of these conditions (such as all actors having wonderful information and no shopper or vendor having market power) do not hold, and market prices alone do not guarantee economical coordination. Complementary arrange-ments, like careful contracts and joint ventures, are unit sometimes needed to induce producers on the vertical chain to retort to the strain of others in the chain however on final customers. The inadequate and inconsistent supplier of quality agricultural outputs to agro-industries in SSA can be a serious constraint on the expansion and gain of agribusinesses. The matter becomes lots of severe as per capita incomes increase in Africa, and as economies become lots of open, resulting in lots of tight quality demands in national and international markets (Kandeh et al., 2011).

A hanging example is that the case of Shea, a tree-based oilseed produced in West Africa that's a major supply of women's financial gain. Since the mid-2000s, world demand for shea butter (the oil extract from the shea nut) has increased dramatically following the choice of the EU Union to permit up to 5% substitution of shea butter as a cocoa butter equivalent in chocolate and also the increased incorporation of shea butter in women's cosmetics in industrialized countries. At the village level, the initial stage of the shea nut method takes two types—a labor-saving type that involves cooking and burial the nut to allow the outer casing to dete-riorate before decupling and extracting the oil, and lots of labor-intensive methodologies involving boiling and sun-drying the balmy. The cooking and burial methodology, though' it's quicker and easier for the women involved, it produces oil containing polycyclic aromatic hydrocarbons (PAHs) that are proverbial carcinogens. As a result of scale economies in assortment and method, nuts processed using both techniques are usually pooled at the village level, with patrons paying a standardized price for the entire heap. The result is that the larger several balmy and butter sometimes contain PAHs, which could finish in rejection of the entire heap for export, forcing it to be sold-out on lots of less profitable domestic market. The absence of clear grades and standards and village-level testing technologies have therefore resulted in many women being excluded from the shea boom—a clear vertical coordination failure. In response, Ghana, and Burkina Faso have developed programmed and created worth chain informative methods involving stakeholders throughout the chain to plug improved process methods at the farm level along with grades, standards, and rating schedules that reward women for the enterprise the lots of time-intensive primary method procedures (Perakis, 2009).

Similar control problems are ubiquitous at intervals the agro-food system and become increasingly necessary as demand in agro-food markets shifts from generic commodities to merchandise that incorporate specific collections of attributes, like food safety and environmental friendliness. Meeting such demands desires lots of tighter vertical coordination heavy chains than sometimes are usually assured by open spot markets. A vital challenge is to vogue institutional arrangements (subcontracting and strategic alliances between farmer organizations and processors) for such coordination that offer an alternative to relying entirely on vertical coordination of big segments of the price chain through possession by a one giant firm, which might limit the participation of the poor in roles apart from low-paid wage labor (Vorley et al., 2009).

2.9.4 ACCESS TO MARKETS

Through the late 1980s, many of the diagnoses of the challenges of agro-food system development in Africa targeted on supply-side constraints. The agricultural market reforms that began at intervals the late 1980s, diode to a gradual shift in stress towards larger recognition that such development needs to be demand-driven, with the matter of however way to link producers throughout the agro-food system to remunerative markets receiving raised attention. A second, newer, shift has been increasing recognition of the importance of regional markets in Africa as a primary offer of demand growth for food merchandise, in distinction to the nearly exclusive focus at intervals the 90s on exports valuable additional food merchandise to overseas markets.

- **Access to International Markets:** Still, access to international markets, notably for high-value merchandise like the farming product and processed product, remains necessary to the event of the enlargement of the agro-food system in SSA. Traditionally, lots of attention has targeted on the matter of tariff increase at intervals the Organization for Economic Cooperation and Development (OECD) countries that produces it more durable for agro-processors to contend in those markets, and on the potential use of photo sanitary standards by high-income countries as a disguised variety of protection (Nouve et al., 2002). Today, however, the foremost necessary barrier to plug access needs to satisfy the standard and traceability standards established by non-public commerce firms (frequently massive retailers) in high-income countries, notably for SMEs. Consequently, it's essential to develop ways to change African agro-food enterprises of all sizes, to participate in world value chains (Vorley et al., 2009).

- **Regional and National Markets:** Diao et al. (2007) showed that the price of regional (intra-African) exports of agricultural merchandise (1996–2000) were quite three times those of exports to non-African markets whereas regional exports, notably of staples, were doable to become the foremost vital for single offer of demand growth for African agro-foods over ensuing 20 years. Accessing such regional markets is less complicated for SSA firms than competitive in overseas markets as a result of the provision demands are less exacting—there is no should be compelled to develop pricey air freight facilities—while quality standards are less rigorous. More-over, national, and regional markets not solely offer companies with doubtless profitable outlets, but to boot operate coaching job grounds, during which they will upgrade their operations to eventually burgled international markets for high-value merchandise.

Unfortunately, access to regional markets is frequently hindered by a range of barriers including:

- High transport prices as a result of poor road infrastructure, super-annuated truck age fleets and railroad instrumentality, high fuel costs and poor storage facilities, also as cold chains. In the late 90s, it cost $230 per ton for meat shipped cows from the Sahel to the West African coast, compared with solely $80 per ton for beef shipped by non-African exporters from the world market (Yade et al., 1999).
- Inadequate data on quantities, qualities, and costs of merchandise offered in neighboring countries, what is more as contact data regarding reliable suppliers. This lack of information, compared to relatively fast access to knowledge regarding overseas suppliers, acts as a non-tariff import barrier, inducing importers in African countries to favor non-African sources of the offer.
- Unreliable systems of law social control and dispute judgment, notably once mercantilism across national borders. The event of regional business enterprise hot organizations offers the prob-ability of exploitation these to develop lots of reliable tools (e.g., personal dispute resolution services) to influence these issues.
- Varied non-official barriers, like roadblocks and bribes to deliver administrative forms, additional raise transport costs, increase uncertainty and contradict government commitments to the free movement of product and folk among regional economic commu-nities like ECOWAS and the Common Market for Eastern and Southern Africa (COMESA).

The results are live lower levels of trade and reduced incentives to intensify production. As an example, Boughton and Dembélé (2010) documented the unofficial charges that traders exporting maize from the Republic of Mali to the Republic of Senegal had to pay, in spite of each countries being members of the ECOWAS trade zone. The charges were similar to the costs for 50 weight unit bags of fertilizer at industrial (unsubsidized) costs for every hectare of maize exported—at a time once the Malian government was subsidizing fertilizers at the rate of 50% to encourage intensification of production.

Without a discount on these restrictions on native and regional market access, African agro-food enterprises will keep disadvantaged relative to foreign competitors, even at markets in Africa (Kandeh et al., 2011).

2.9.5   MANAGERIAL CAPACITY

For the tiniest agribusinesses-household-level, microenterprises-management constraints cause a severe limit on growth. Of times, the 'border' between such enterprises and additionally the social unit is blurred, with no notions of separate firm accounting, employment, or financial flows. Access to data on technologies and promising market retailers is commonly restricted as a result of the illiteracy of the entrepreneurs. For SMEs, firm management challenges revolve around distinctive and adapting to promising markets and technologies, developing agreement relationships with larger actors up and down the worth chain and accessing finance. The foremost vital agro-industrial firms are of times involved in several levels of the worth chain, and so a vital a component of their management attention is targeted on developing improved vertical coordination arrangements, to assure the reliability and quality of their inputs and outputs (Kandeh et al., 2011).

2.9.6   ACCESS TO FINANCE

Given scarce public resources, what mixture of firms should public policies and investments try to foster? The answer depends on the potential of assorted sized firms to grow and re-create wealth and remunerative employment that in turn depends on a decent several of components, like scale and agglomeration economies, ability to reply to chop-chop dynamic demands, and additionally the linkages fully at intervals firms generate with the rest of the economy. Certainly, there is scope for large-scale enterprises in some areas, considerably in light-weight of growing foreign direct investment (FDI) in SSA, increasingly from Asia and aimed toward serving growing markets thereon continent (Broadman et al., 2007). For these sorts of firms, clear foreign investment codes and alternative components of the facultative atmosphere are crucial. Turning out with restrictive frameworks that build it easier for them to partner with farmer associations and totally different suppliers may additionally be very important in serving to assure that these enterprises profit a broad vary of actors at intervals the economy, rather than being beholden to arrange all their activities internally through their own utilized labor. Notably, policies that by artificial means depress the value of capital could also be prejudices; in this, they will induce premature adoption of labor-saving instrumentality at a time once Africa needs to generate lots of jobs for its burgeoning labor force.

At the alternative end of the spectrum are tiny firms, sometimes part-time family operations familiarized to terribly a really awfully native client base and generating very low levels of financial gain. Rather like the littlest farms, these enterprises are a component of poor families' survival strategies; but their growth prospects over the medium term are slim. It's visiting be a less complicated use of public resources to foster growth in numerous segments of agroindustry, therefore, tax variety of these earnings to finance programs that equip the house owners of the microenterprises with the tools to maneuver into higher-earning opportunities at intervals the labor market—e.g., through operational for the medium or large-scale agro-enterprises. American State Jeanery (2009) discusses how such a transition to wage labor raised the incomes of former small-scale farmers who shifted to figure for inexperienced bean commerce firms in the Republic of Senegal. Between these two extremes lie the SMEs, variety of that has substantial growth potential, considerably to serve growing national and regional markets; however which regularly would like technical, financial, informational, and structure help to beat the constraints mentioned on top of (Kandeh et al., 2011).

2.10 OVERCOMING CONSTRAINTS TO FOSTER GROWTH

Agro-industries in SSA necessity to endure a structural transformation as profound as that required of farming over ensuing 20 years, therefore on get the roles, incomes, and food merchandise thus badly every needed by Africa's growing population. Indeed, the transformations of farming and agro-industry are inextricably connected, and the growth of vivacious agro-industries is important to supply employment for an oversized range of husbandman farmers who are unlikely to farm their way out of poverty.

Rapidly renascent demands and technologies mean that agro-industries in Africa should be nimble if they are to be competitive with producers from different regions of the world. The good news is that the policy setting and so the technological tools needed to foster the transformation of the business enterprise system is the foremost favorable they have been in over a generation. Growing external and internal demand offers likely profitable retailers for agro-industrial merchandise, if the problems of market access connected to high transport and event prices, also because the costs of meeting increasingly exigent quality standards could also be overcome. African governments and their development partners should concentrate on key investments and policy changes that 'crowd in' investment by non-public actors and open a political area for autonomous business organizations to work to resolve the issues of horizontal and vertical coordination that presently constrain the expansion in agro-industry. An adequate offer of agricultural raw materials for the agro-processing business is very important for achieving productivity growth in markets wherever agro-industrial value chains predominate; highlight the need to upgrade business enterprise value chains (Kandeh et al., 2011).

2.11 CONCLUSION

The worldwide community has diagnosed the demanding situations and the need for trans-formative change. In particular, the 2030 Agenda for Sustainable Development, adopted with the aid of the worldwide community in September 2015, provides a compelling, however challenging, imaginative and prescient on how a couple of goals can be combined to outline new sustainable development pathways. The 2nd Sustainable Development Goal (SDG 2) is explicitly ambitions at finishing hunger, accomplishing food protection and improved nutrients, and promoting sustainable agriculture, simultaneously by 2030. Where the 2030 Agenda recognizes that development closer to many different SDGs, in particular the eradication of poverty and the response to weather change (SDG 13) and the sustainable use of marine and terrestrial ecosystems (SDG 14 and 15), will rely on the extent to which meals insecurity and malnutrition are efficaciously decreased and sustainable agriculture is promoted. Conversely, development closer to SDG 2 will depend upon development made towards several of the other goals. In different words, in an effort to make progress on SDG 2, policy-makers and all different stakeholders will want to consider interlink ages and critical interactions, both in phrases of synergies and trade-offs, between SDG 2 and all different goals.

The 2030 Agenda for Sustainable Development and the Addis Ababa Action Agenda on financing for development specifically call on all nations to pursue policy coherence and set up permitting environments for sustainable development at all ranges and by using all actors (SDG 17). The Paris Agreement on climate change and the steps closer to its implementation taken on the United Nations Climate Change Conference 2016 (COP22) in Marrakesh, mirror global commitments for concerted action to cope with the perils of weather change. The Sendai Framework for Disaster Risk Reduction also offers priority to the agriculture sectors. Despite those promising worldwide frameworks for movement, reaching policy coherence may be challenging. The 2030 Agenda and different related international agreements pressure the interdependence of the challenges they may be to address. They also understand the want to integrate exclusive actions to reap linked objectives and that doing so will pose new technical needs on policy-makers, at all ranges, in addition to new needs on institutional arrangements and coordination at numerous tiers of governance. The associated demanding situations are twofold. First, distinctive gadgets applied at distinct tiers of governance will want to be mixed in ways which might be mutually reinforcing, even as inevitable trade-offs are identified and contained. Second, capitalizing on synergies among SDGs and targets, among exceptional sect-oral policies, and among diverse moves undertaken by using officers and stakeholders at stages that range from local, municipal, and provincial to countrywide, and from national to regional and international, has proven pretty challenging in the past.

The purpose of this book is not to provide a menu of solutions, however rather to increase know-how of the nature of the challenges that agriculture, rural development and food structures are facing now and will be dealing with into the 21st century. The evaluation presented here of global trends and demanding situations gives similarly insights into what's at stake and what wishes to be done. The following chapters assess developments that will form the destiny of meals and the livelihoods of those relying on food and agricultural systems. Most of the developments are strongly interdependent and, blended, inform a set of challenges to reaching food security and nutrition for all and akin agriculture sustainable. One clear message that emerges is that 'business-as-usual' isn't an option. Major alterations of agricultural structures, rural economies and natural resource management will be wished if we are to satisfy the more than one challenge before us and recognize the full ability of food and agriculture to make sure a secure and healthful destiny for all of us and the complete planet.

KEYWORDS

- **agribusiness**
- **agricultural sector**
- **food security**
- **International Finance Corporation (IFC)**
- **poverty reduction**
- **sub-Saharan Africa (SSA)**

REFERENCES

American State Jeanery, (2009). *Free, Easy to Search Official Information About Every UK Company Registered with Companies House*. https://www.thegazette.co.uk (accessed on 4 January 2020).

Amit, R., (2009). *Global Fertilizer Situation and Fertilizer Access*. Presentation to the World Bank's 2009. Agriculture and Rural Development Week. Available at: http://siteresources.worldbank.org/INTARD/Resources/335807–1236361651968/ARDmeetingpresentation_WorldBank_March32009.pdf (accessed on 4 January 2020).

Boehlje, M., & Schrader, L. F., (1998). The industrialization of agriculture: Questions of coordination. In: Royer, J. S., & Rogers, R. T., (eds.), *The Industrialization of Agriculture: Vertical Coordination in the US Food System* (pp. 3–26). Brookfield, Vermont: Ashgate.

Boughton, D., & Dembélé, N. N., (2010). *Rapid Reconnaissance of Coarse Grain Production and Marketing in the CMDT Zone of Southern Mali: Field Work Report of the IER-CSAPROMISAM Team*. PRESAO Working Paper no. 2010–1. Bamako: Michigan State University.

Boughton, D., Crawford, C., Howard, J., Oehmke, J., Shaffer, J., & Staatz, J., (1995). *A Strategic Approach to Agricultural Research Program Planning in Sub-Saharan Africa*. MSU International Development Working Paper no. 49. East Lansing, MI: Michigan State University, Department of Agricultural Economics.

Broadman, H. G., Isik, G., Plaza, S., Ye, X., & Yoshino, Y., (2007). *Africa's Silk Road: China and India's New Economic Frontier*. Washington D.C.: The World Bank.

Brown, Casey, Robyn Meeks, Yonas Ghile, & Kenneth Hunu (2013). Is water security necessary? An empirical analysis of the effects of climate hazards on national-level economic growth, *Philosophical Transactions of the Royal Society, A, 371*, [Google Scholar] [Crossruff].

Brown, C., Meeks, R., Hunu, K., & Yu, W., (2011). Hydroclimate risk to economic growth in sub-Saharan Africa. *Climatic Change, 106*, 621–647.

Burke, E. J., & Brown, S. J., (2008). Evaluating uncertainties in the projection of future drought. *Journal of Hydrometeorology, 9*(2), 292–299.

Burke, E. J., Brown, S. J., & Christidis, N., (2006). Modeling the recent evolution of global drought and projections for the 21st century with the Hadley center climate model. *Journal of Hydrometeorology, 7,* 1113–1125.

Charmes, J., (2000). *African Women in Food Processing: A Major, But Still Underestimated Sector of Their Contribution to the National Economy.* Paper Prepared for the International Development Research Centre (IDRC), Ottawa: IDRC.

Christy, R., Mabaya, E., Wilson, N., Mutambatsere, E., & Mhlanga, N., (2009). Enabling environments for competitive Agro-industries. In: Da Silva, C., Baker, D., Shepherd, A. W., Jenane, C., & Miranda-Da-Cruz, S., (eds.), *Agro-Industries for Development* (pp. 136–185). Wallingford, UK: CABI for FAO and UNIDO.

Dennis, C., Aguilera, J., & Satin, M., (2009). Technologies shaping the future. In: Da Silva, C., Baker, D., Shepherd, A. W., Jenane, C., & Miranda-da-Cruz, S., (eds.), *Agro-Industries for Development* (pp. 92–135). Wallingford, UK: CABI for FAO and UNIDO.

Diao, X., Hazell, P., Resnick, D., & Hurlow, J., (2007). *The Role of Agriculture in Development: Implications for Sub-Saharan Africa.* IFPRI Research Report 153, Washington D.C.: IFPRI.

Diarra, S., Staatz, J., & Dembélé, N., (2000). The reform of rice milling and marketing in the Office du Niger: Catalysts for an agricultural success story in Mali. In: James, B. R., Robinson, D., & Staatz, J., (eds.), *Democracy and Development in Mali* (pp. 167–188). East Lansing: Michigan State University.

ECOWAS, (2005). Economic Community of West African States. The regional approach to biosafety in West Africa. In: *ECOWAS Ministerial Conference on Biotechnology: Strategies and Actions for Sustainable Agricultural Production: Safe for Humans and the Environment,* Bamako.

ECOWAS, (2009). *Economic Community of West African States: Regional Agricultural Policy for West Africa: ECOWAP/CAADP.* Document presented at the international conference on financing regional agricultural policy in West Africa (ECOWAP/CAADP). Abuja: ECOWAS.

Eicher, C., Maredia, K., & Sithole-Niang, I., (2006). Crop biotechnology and the African farmer. *Food Policy, 31*(6), 504–527.

FAO, (2007). Food and Agriculture Organization of the United Nations. *Challenges of Agribusiness and Agro-Industry Development.* Available at: http://ftp.fao.org/docrep/fao/meeting/011/j9176e.pdf (accessed on 16 January 2020).

Ferré, T., Doassem, J., & Kameni, A. (1999). Dynamics of agricultural product processing activities in Garoua, North Cameroon, Garoua, Cameroon: Institute for Agricultural Research for Development (IRAD)/Pole of Research Applied to Development of Savannas of Central Africa (PRASAC).

GRAIN, (2008). *Seized: The 2008 Land Grab for Food and Financial Security.* Briefing. Barcelona: Genetic Resources Action International (GRAIN). Available at: http://www.grain.org/briefings_files/landgrab-2008-en.pdf (accessed on 4 January 2020).

Hamada, Y. M. (2018). Special pictures album, Took by Youssef M. Hamada, Egypt, 2018.

Heisey, P., & Mwangi, W., (1997). Fertilizer use and maize production in Sub-Saharan Africa. In: Byerlee, D., & Eicher, C. K., (eds.), *Africa's Emerging Maize Revolution.* Boulder, Colorado: Lynne Rienner Publishers.

IFC, (2014). *International Finance Corporation.* South Asia. www.ifc.org (accessed on 16 January 2020).

Ilboudo, S., & Kambou, D. (2009). *Development of An Operational Agro-Industry Development Strategy in the Sahel and in West Africa*: Provisional report. Ouagadougou: CILSS Executive Secretariat Regional Support Program/Market Access.

Jaffee, S., Kopicki, R., Labaste, P., & Christie, I., (2003). *Modernizing Africa's Agro-Food Systems, Analytical Framework and Implications for Operations*. Africa Region Working Paper Series No. 44, Washington D.C.: The World Bank.

Jayne, T. S., & Jones, S., (1997). Food marketing and pricing policy in eastern and southern Africa: A survey. *World Development, 25*(9), 1505–1527.

Kandeh, K. Y., Patrick, M. K., Torben, M. R., & Anthony, M. H., (2011). *Agribusiness for Africa's Prosperity*. UNIDO ID/440, Layout by Smith + Bell Design (UK), Printed in Austria.

Lenaghan, T., (2009). *Global Food Security Response: Mali Rice Study*. Washington D.C.: United States Agency for International Development (USAID).

Malcolm, S., Marshall, E., Aillery, M., Heisey, P., Livingston, M., & Day-Rubenstein, K., (2012). In: *Agricultural Adaptation to a Changing Climate: Economic and Environmental Implications Vary by U.S. Region* (p. 84). Economic Research Report No. (ERR-136). http://www.ers.usda.gov/media/848748/err136.pdf (accessed on 16 January 2020).

Me-Nsope, N., Staatz, J., & Dembélé, N. (2008). Literature review on the role of grain grades and standards in promoting regional agricultural trade: Implications for West Africa. In: PRESAO Working Papers. Bamako: Research and Capacity Building Program in Food Security in West Africa.

Montalvo, J., & Ravallion, M., (2010). The pattern of growth and poverty reduction in China. *Journal of Comparative Economics, 38*(1), 2–16.

Morris, M., Kelly, V. A., Kopicki, R. J., & Byerlee, D., (2007). *Fertilizer Use in African Agriculture: Lessons Learned and Good Practice Guidelines*. Washington D.C.: The World Bank.

Muchnik, J., (2003). *Food, Know-How and Agrifood Innovations in West Africa*, Collection of reports from the ALISA project, European Union, DG XII, Brussels.

Nagai, T., (2008). *Competitiveness of Cowpea-Based Processed Products: A Case Study in Ghana*. A thesis submitted to Michigan State University.

Nienke, M., & Stads, G., (2006). *Agricultural R&D in Sub-Saharan Africa: An Era of Stagnation*. Background Report for Agricultural Science and Technology Indicators (ASTI) Initiative Washington D.C.: IFPRI. Available at: http://www.asti.cgiar.org/pdf/AfricaRpt_200608.pdf (accessed on 16 January 2020).

Nouve, K., Staatz, J., Schweikhardt, D., & Yade, M., (2002). *Trading out of Poverty: WTO Agreements and the West African Agriculture*. MSU International Development Working Paper No 80. Michigan: Michigan State University Department of Agricultural Economics.

Pardey, P. G., Beintema, N., Dehmer, S., & Wood, S., (2006). Agricultural research: A growing global divide?. In: *Agricultural Science and Technology Indicators Initiative*. Washington D.C.: IFPRI.

Perakis, S., (2009). *Improving the Quality of Women's Gold in Mali*. West Africa: The Case of Shea. Michigan State University.

Ringler, C., Zhu, T., Gruber, S., Treguer, R., Laurent, A., Addams, L., Cenacchi, N., & Sulser, T., (2016). Role of water security for agricultural and economic development: Concepts and global scenarios. In: Claudia, P. W., Anik, B., & Joyeeta, G., (eds.),

Handbook on Water Security (pp. 183–200). Chapter 11. Cheltenham, UK: Edward Elgar Publishing Limited.

Rosegrant, M. W., (2015). Global outlook for water scarcity, food security, and hydropower. In: Kimberly, B., Richard, H., James, A. R., & Christopher, A. W., (eds.), *Handbook of Water Economics and Institutions*. Chapter 1. New York, NY, USA: Rutledge.

Rosegrant, M. W., (2016). *Challenges and Policies for Global Water and Food Security.* Agriculture's Water Economy Symposium. Federal Reserve Bank of Kansas City, Kansas City, Missouri, USA.

Sadoff, C. W., Hall, J. W., Grey, D., Aerts, J. C. J. H., Ait-Kadi, M., et al., (2015). *Securing Water, Sustaining Growth: Report of the GWP/OECD Task Force on Water Security and Sustainable Growth* (p. 180). University of Oxford, UK.

Sahel Institute, (2010). *Sahel Institute Database.* Available at: http://196.200.57.138/dbinsah/index.cfm?sect1=pesticide&id=58&quer=pesticide1.cfm&sect2=x (accessed on 16 January 2020).

Sautier, D., Vermeulen, H., Fok, M., & Biénabe, E., (2006*). Case Studies of Agri-Processing and Contract Agriculture in Africa.* RIMISP, Latin American Center for Rural Development, Santiago: RIMISP.

Toe, A. M., (2009). *Pesticide Registration Process in CILSS Countries.* Presentation to the Global MRL Harmonization Initiative in Africa, Alexandra, Egypt.

Von, B. J., & Meinzen-Dick, R., (2009). *Land Grabbing by Foreign Investors in Developing Countries: Risks and Opportunities.* IFPRI Policy Brief, 13, Washington, D.C.: IFPRI.

Vorley, B., Lundy, M., & MacGregor, J., (2009). Business models that are inclusive of small farmers. In: Da Silva, C., Baker, D., Shepherd, A. W., Jenane, C., & Miranda-da-Cruz, S., (eds.), *Agro-Industries for Development* (pp. 186–222). Wallingford, UK: CABI for FAO and UNIDO.

Wilkinson, J., & Rocha, R., (2008). *Agro-Industries, Trends, Patterns and Developmental Impacts.* Paper prepared for global agro-industries forum (GAIF), New Delhi.

World Bank, (2007). *World Development Report 2008: Agriculture for Development.* Washington, D.C.

World Bank, (2009). *World Development Indicators 2009.* Washington D.C.: The World Bank.

World Bank, (2010). World Development Report. Chapter 3, Managing land and water to feed nine billion people and protect natural systems. In: *WDR 2010, Development and Climate Change.* World Bank, Washington DC, USA. http://siteresources.worldbank.org/INTWDR2010/Resources/5287678–1226014527953/Chapter-3.pdf (accessed on 4 January 2020).

Yade, M., Chohin-Kuper, A., Kelly, V., Staatz, J., & Tefft, J., (1999). *The Role of Regional Trade in Agricultural Transformation: The Case of West Africa Following the Devaluation of the CFA Franc.* Paper presented at the workshop on agricultural transformation. Sponsored by Tegemeo Institute/Egerton University, Njoro, Kenya; Eastern and Central Africa Programme for Agricultural Policy Analysis (ECAPAPA), Entebbe, Uganda; Michigan State University, East Lansing, Michigan, USA; and the United States Agency for International Development (USAID).

CHAPTER 3

Climate Change: Causes and Effects

ABSTRACT

The world is on an important interval in its exertions towards the struggling world of global climate change. Since the first meeting of the Parties in 1995, greenhouse-gas (GHG) emissions have grown up by quite one-quarter and conjointly the filled with atmosphere concentration of these gases has inflated a lot of and a lot of to 435 elements per million carbon dioxide equivalent (ppm CO_2-eq) in 2012 (EEA, 2015).

The Global Panel on world global climate change has established that, within the absence of fully committed and crucial action, world climate change will have severe and irreversible effects throughout the world. The international commitment plenty to save lots of the increase in long traditional heats to below 2°C, relative to pre-industrial levels, would need extended and constant decreases in world emissions. Figure 3.1 shows the Nile on a hot day in Upper Egypt, Egypt.

3.1 BACKGROUND

Development of civilization would haven't happened whereas not carbon emission in most cases. But carbon emissions are presently believed to feature to warming and consequent world global climate change events. Scientists believe that the planet has already burnt half the fossil fuels necessary to cause a 2°C rise in world temperature (Chandrappa et al., 2011). Because the humans began to settle their energy demand put together inflated with wood as the main offer of energy. By the 1280s, folks started using coal for fuel in processes like line kilns and shaping that resulted in pollution having black smoke and oxides of sulphureted in its emissions. Late 18th and early 19th centuries witnessed major changes in agriculture, producing, production, mining, and transportation. The onset

FIGURE 3.1   The Nile on a hot day in Upper Egypt, Egypt (Hamada, 2018).

of the industrial revolution marked a turning purpose for world global climate change. The employment of fuel in street lighting was eventually replaced with the emergence of the trendy electrical era. With the event of power among the 19th century, coal's future became closely tied to electricity generation. The first wise coal laid-off electrical generating station, developed by Thomas Alva Edison, went into operation in New York in 1882, offer electricity for unit lights (Chandrappa et al., 2011). Oil overtook coal as a result of the most important offer of primary energy among the 60s, with the huge growth among the transportation sector. Coal still plays a major role among the world's primary energy mix, providing 23.5% of world primary energy and 39th of the world's electricity in 2002 (Chandrappa et al., 2011).

The industrial revolution had a wonderful result on the socioeconomic and cultural conditions were starting in the United Kingdom, followed by

Europe, North America, and eventually the world. The economic revolution marks a major turning purpose in human history. Starting among the later a component of the 18th century, nice Britain's previously toil and animal-based economy modified to machine-based manufacturing. It started with the mechanization of the textile industries followed by the development of iron-making techniques that lead to increased use of coal. The developments of machine tools are among the initial 20 years of the 19th-century diode to producing lots of machines for various industries. The first age, that began among the 18th-century diode to second age in the 19th century, with the event of powered ships and railways. Nineteenth-century witnessed combustion engine and electrical power generation. With the industrial revolution came a series of environmental impacts—pollution, water pollution, thermal pollution, noise pollution and degradation of forest and different ecosystems. Increase in carbonic acid gas semiconductor diode to heating as a result of physical phenomenon (Chandrappa et al., 2011).

3.2 SCOPE OF THE CHAPTER

Climate change may be a true and imperative challenge that is already poignant and also the environment worldwide. Very important changes are occurring on Earth, along with increasing air and ocean temperatures, widespread melting of snow and ice, and rising ocean levels. This chapter discusses key scientific facts that designate the causes and effects of global climate change today, conjointly as projections for the long run (EPA, 2010).

Recent observations of warming support the thought that greenhouse gases are warming the world. Over the last century, the world has experienced the foremost necessary increase in surface temperature in one, 300 years. The common surface temperature of the planet rose 0.6 to 0.9°C (1.08°F to 1.62°F) between 1906 and 2006, and conjointly the speed of temperature increase nearly doubled among the last 50 years. Worldwide measurements of water level show a rise of 0.17 meters (0.56 feet) throughout the 20th century. The world's glaciers have steadily receded, and the Arctic Ocean ice extent has steadily shrunken by 2.7% per decade since 1978. While greenhouse emission concentrations stable today, the world would still heat by regarding (0.6°C) over a serial century as a result of it takes years for Earth to altogether react to an increase in greenhouse gases. As Earth has heat, lots of the excess energy has gone into heating the upper layers of the ocean. Scientists suspect that currents have transported a variety of this excess heat from surface waters down deep, removing it from the surface of our planet. Once the lower layers of the ocean have heat, the excess heat among the upper layers will no longer be drawn down, and Earth will heat regarding 0.6°C (1°F) (Riebeek, 2007). The aim of this chapter is to debate facts that designate the causes and effects of global climate change today.

3.3   GENERAL OVERVIEW OF CLIMATE CHANGE

Global warming and world climate change are essentially attributed to the emission of GHGs from natural or phylogenies' sources and changes in magnitude relation, a reflection of radiation from altogether different surfaces to the atmosphere that causes warming or cooling of the planet, with degree between 0–1 (Wikipedia, 2012). Amendment global climate change is one of the foremost drivers of terrestrial natural phenomenon change and has altogether different effects, like disturbances and loss of environment. To boot, following an amendment in climate parameters (precipitation variation, snow cover, humidity, sea level, etc.) there is variation in the exchange of assorted activities in dependent, prediction, parasitic, and mutualistic relationships (Lepetz et al., 2009). As international mean temperature rises, it causes positive or negative effects on altogether completely different processes and activities in earth systems (IPCC, 2007).

The effects of world global climate change became obvious among the natural atmosphere over the last 30 years, aboard completely different threats like environment destruction, fragmentation, disturbance, and loss in diversity (Lepetz et al., 2009). For example, the land use amendment (the most important impact) in tropical forests can cause loss of diversity. Hence, exploitation of natural resources, use of hardwood timber and forest clearing causes high loss among the amount and accessibility of habitats, and to the extinction of species, significantly that are vulnerable, and restricted in vary. Although it's hard to make a contributing link between amendments of climate in reference to change in species richness, as a results of many completely different variables are also involved (Morris, 2010), species can act every directly and indirectly and in most cases, indirect interaction is unpredictable (Yodzis, 2000; Montoya et al., 2005; Morris, 2010). World global climate change affects species indirectly by reducing the amount and accessibility of habitats and by eliminating species that unit of measurement essential to the species in question (Morris, 2010). As a result, the loss of one species might lead to a decrease, increase, or extinction of various apparently unconnected species; however, the human activities unit of measurement inflicting secondary extinction at a higher level than expected from random species losses. Once species unit of measurement lost from a system, it isn't the only real species that is lost, but the interaction and conjointly the overall ecological functions, that we tend to expect from these interactions, are going to be conjointly lost (Morris, 2010).

According to FAO (2007) and Mimura (2010) global climate change impacts classified into two broad categories:

1. Biophysical impacts: indicates the physical impacts caused by climate change directly in physical environment; example, drought, and flooding, causes an effect on physical surroundings like: (a) effects on quality and quantity of crops, pasture, forest, and farm animal; (b) Modification in natural resources quality and amount of soil, land, and water resources; (c) Enlarged weed and insect cuss challenges as a results of world climate change; (d) Shift in special and temporal distribution of impacts, (sea level rise, modification in ocean salinity, and ocean temperature that rise inflicting fish to inhabit totally different ranges).
2. Socio economic impacts: following the first biophysical impacts on surroundings there will be a secondary result on socio economic systems. For example, decline in yield, reduced marginal gross domestic product from agriculture sector, fluctuation of world value, amendment in geographical distribution of trade regimes, as a results of shortage of food in quality and quantity the number of individuals in hunger and risk enlarged, and cause migration.

3.4 A SIMPLIFIED GUIDE TO THE IPCC'S CLIMATE CHANGE 2001: IMPACTS, ADAPTATION, AND VULNERABILITY

Global warming is already changing the world around us in ways in which within researchers cannot measure it them carefully. Such changes will become lots of evident with each passing decade. Although cutting greenhouse emissions therefore minimize future global climate change ought to be our high priority, we've a bent to ought to put together prepare to reply to impacts that our past emissions presently produce inevitable. Folks everywhere have got to be compelled to understand that global climate change goes to have an effect on them and think what they're going to do to cope. Luckily, working party II of the WMO/UNEP Intergovernmental Panel on climate change (IPCC) has assessed what researchers have learned concerning expected impacts and therefore the think to adapt to them.

By findings are presented in an exceedingly comprehensive publication entitled "Climate modification 2001: Impacts, Adaptation and Vulnerability," that's part of the IPCC's Third Assessment Report. This Report put together includes volumes on the causes of global climate change and on decisions for limiting greenhouse emissions. This simplified guide presents the very technical findings of working party II in everyday language. It isn't a political candidate document and has been neither approved nor accepted by the IPCC. Instead, it seeks to form the various pages of careful text contained how the "Impacts, Adaptation and Vulnerability" lots of accessible to a broader audience. Readers must be galvanized to hunt a lot of data by consult with the primary publications and thus the IPCC's electronic computer (Töpfer, 2003).

3.5   EFFECTS OF CLIMATE CHANGE

Look closely and you'll see the results of global climate change. Scientists have documented climate-induced changes in some 100 physical and 450 biological processes. Among the Russian Arctic, higher temperatures are melting the land, inflicting the foundations of five-story flat buildings to slump. Worldwide, the rain, once it falls, is typically lots of intense. Floods and storms are lots of severe, and heat waves are becoming lots of utmost. Rivers freeze later among the winter and soften earlier. Trees flower earlier in spring, insects emerge faster, and birds lay eggs sooner. Glaciers are melting. The global mean sea level is rising.

3.6   EARTH'S CLIMATE HAS CHANGED OVER THE PAST CENTURY

Earth's climate has modified over the past century. The atmosphere and oceans heat, ocean levels have up, and glaciers and ice sheets have reduced in size. The best accessible proof indicates that greenhouse emission emissions from human activities are the foremost cause. Continued can increase in greenhouse gases will turn out further warming and completely different changes in Earth's physical setting and ecosystems (Holmes, 2015).

3.7 WHAT IS CLIMATE CHANGE?

The term 'climate,' in its broadest sense, refers to a mathematical description of weather and of the connected conditions of oceans, land surfaces, and ice sheets. This includes thought of averages, variability, and extremes. Global climate change is an associate in alteration among the pattern of climate over an extended quantity of our time, and will ensure to a mixture of natural and human-induced causes (Holmes, 2015).

3.8 HOW HAS CLIMATE CHANGED?

Global climate has varied greatly throughout Earth's history. Among the ultimate decades of the 20th century, the world seasoned a rate of warming that is unprecedented for thousands of years, as most as we are going to tell from the on the market proof. International average temperature rise has been among current rises in ocean temperatures, ocean heat storage, ocean levels, and atmospherically vapor. There has additionally been shrinkage among the scale of ice sheets and most glaciers. The recent delay among the speed of surface warming is particularly as a result of climate variability that has localized heat among the ocean, inflicting warming at depth, and cooling of surface waters. Australia's climate has heat in conjunction with the worldwide average warming (Holmes, 2015).

3.9 WHAT DOES SCIENCE SAY ABOUT OPTIONS TO ADDRESS CLIMATE CHANGE?

Societies, along with Australia, face decisions regarding how to reply to the implications of future global climate change. Accessible ways embody reducing emissions, capturing carbonic acid gas, adaptation, and 'geo-engineering.' These ways, which could be combined to some extent, carry fully different levels of environmental risk and different group consequences. The role of climate science is to inform picks by providing the best potential information on climate outcomes and conjointly the implications of alternative courses of action (Holmes, 2015).

3.10 MANY UNCERTAINTIES ARE LIKELY TO PERSIST

A number of things stop lots of correct predictions of world climate change, and many of these will persist. Whereas advances still are created in our understanding of climate physics and thus the response of the climate system to an increase in greenhouse gases, many uncertainties are most likely to persist. The speed of future warming depends on future emissions, feedback processes that dampen or reinforce disturbances to the climate system, and unpredictable natural influences on climate like volcanic eruptions. Unsure processes that will have an impact on however briskly the world warms for a given emissions pathway are life dominated by cloud formation (Stevens and Bony, 2013), but put together embrace vapor and ice feedbacks, ocean circulation changes, and natural cycles of greenhouse gases.

Although datagram past climate changes for the foremost half corroborates model calculations, typically this can be often put together unsure as a result of inaccuracies among info the data and possibly very important factors regarding that we have incomplete information. It's really hard to tell intimately but world global climate change will have an impact on individual locations, notably with relevance precipitation. While a worldwide change was loosely explanted, its regional expression would depend on elaborate changes in wind patterns, ocean currents, plants, and soils. The climate system can gift abruptness: abrupt climate transitions have occurred in Earth's history, the temporal configuration and probability of that can't usually be foretold with confidence.

3.11 DESPITE THESE UNCERTAINTIES

Despite these uncertainties, there is near-unanimous agreement among climate scientists that human-caused warming is real (Doran et al., 2014; Ander Egg et al., 2010; Cook et al., 2013). It's explanted that human activities since the industrial revolution have sharply inflated gas concentrations; these gases have a warming effect; warming has been determined; the calculated warming is amounting to the discovered warming.

Continued reliance on fossil fuels would cause larger impacts among the long run than if this were curtailed. This understanding represents the work of thousands of specialists over a century, and is awfully unlikely to be altered by further discoveries.

3.12 CAUSES OF CLIMATE CHANGE

Climate change is a term that refers to major changes in temperature, rainfall, snow, or wind patterns lasting for several years or longer. Every human created and natural issue contributes to climate change:

- Human causes embrace burning fossil fuels, deforest, and change agriculture land to cities, and roads.
- Natural causes embrace changes among the Earth's orbit, the sun's intensity, the circulation of the ocean and thus the atmosphere, and volcanic activity.

Although the Earth's climate has changed another time and another time throughout its history, the speedy warming seen today cannot be explained by natural processes alone. Human activities are increasing the quantity of greenhouse gases among the atmosphere. Some amount of greenhouse gases is vital continuously to exist on Earth—they amendment heat among the atmosphere, keeping the planet heat in a very state of equilibrium. But this natural atmospheric phenomenon is being strengthened as human activities (such as the combustion of fossil fuels) add lots of these gases to the atmosphere, resulting in a shift among the Earth's equilibrium (EPA, 2010).

3.13 SIGNS OF CLIMATE CHANGE

Climate change goes on presently, and thus the results are also seen on every continent and in every ocean. Whereas sure effects of action are also helpful, notably among the short term, current, and future effects of action produce big risks to people's health and welfare, and thus the environment.

There is presently clear proof that the Earth's climate is warming (EPA, 2010):

- Land surface temperatures have up by 1.3 (°F) of degrees physicists over the last 100 years.

• Worldwide, the last decade has been the warmest on record.
• The speed of warming across the planet over the last 50 years (0.24°F per decade) is almost double the speed of warming over the last 100 years (0.13°F per decade).

3.14 THE POST KYOTO AGENDA: THE ENERGY FRONTIERS

Burning quite around one-quarter of the economic reserves of oil, coal, and gas, will unleash greenhouse gases to make a major risk of ruinous action. At this rate of fuel use, we have a tendency to be going to consume this quantity in 40 years or less. Despite this case and thus the warnings from the world's leading climate scientists, the industry is embarking on a taxpayer-supported international enlargement into the most reaches of the global: from the ice-bound seas of the Arctic, the deep waters of the northeast Atlantic West Africa and Russia, to Latin America and the tropical waters of northern Australia. In choosing to expand our 1997 campaign challenge to halt new oil exploration and development, the nongovernmental organization seeks to draw a line around the current reserve of fossil fuels and campaign against its further growth. This making known offers a quick factual outline of the oil frontiers, and a prime level read of the oil companies' response to the action. It outlines Greenpeace's principle for targeting new oil frontiers and highlights the contradictions between the oil industry's rhetoric on temperature change and action in continued to expand its exploration for fossil fuels.

In 1998, determinable $94 billion are spent worldwide on oil and gas exploration and production, an increase of $85 billion in 1997. This represents the second-highest growth in disbursement among the last decade and thus the third consecutive year of number growth. It conjointly undermines efforts by the 160 national governments that united to the legally binding Kyoto Protocol (KP), in Japan to chop greenhouse gases. Taxpayers, significantly within US and Europe, are supporting this growth of oil exploration by a number of the world's richest corporations like Shell, Exxon, BP, ARCO, and Chevron. The generality recent analysis shows that in 1995, the U.S. Government, by the foremost conservative estimate, provided $5.2 billion in tax breaks and cash facilitate to the oil sector. In the US, these major oil corporations like Exxon and Shell, have annual revenues larger than the GDP of states like Greece, Venezuela, and New Zealand. The U.S. Government, however, has ablated the rate over the last 20 years, and oil corporations have really paid well less—only one 0.5 to a minimum of one-third of that charge. Whereas BP and Shell have acknowledged the hazards expose by global climate change and therefore the want for preventative action, they have did not mirror this apparent concern in their investments. The dioxide emissions from burning the fossil fuels that Shell created in 1995 are on prime of the dioxide emissions from the complete of North American countries among the identical year. Shell spent around $7.5 billion in 1997 on oil exploration and production. In distinction, the company plans to pay a mean of only $100 million over sequent five years on renewables. BP aimed for $1 billion in star product sales by 2010, with an annual investment of $15 to $20 million to realize this. But typically, this can be often dwarfed by the $4 billion it spent on exploration and production in 1997.

BP and Shell's statements in favor of renewable energy are a welcome success, however, it's clear that the pace of investment in renewables is far too slow to avoid dangerous action, and is being swamped by investments to expand their oil base. In the meantime copious of the rest of the business remains vehemently essential taking action against the action and remains resolute disbursement millions creating an endeavor to convert the overall public there is nothing to worry concerning. The yank fossil fuel Institute, whose members embrace BP, Shell, Exxon, and Mobil, was in May 1998 bobbing up with a $6 million campaign specifically to undermine public confidence in climate science. Unless governments act currently to prevent more exploration to accelerate the transfer of investment into the renewable energy, the planet is committed to new oil, gas, and coal developments that the planet's climate won't be able to withstand (Hamilton, 1998).

3.15   HUMAN ACTIVITIES RESULT IN EMISSIONS

It is regarding 100 years past that scientists raised a priority regarding the burning of fossil fuels throughout the 19th century that they warned would change the climate. Top among these scientists was Savant Arrhenius of Sweden who won the Nobel Peace Prize for Chemistry in 1903. Warnings of those scientists were unnoticed in society's ambition to get pleasure from luxury. Production of cars and their possession in developed countries lead to a further increase in dioxide in the atmosphere. With the dawn of the industrial revolution, the global witnessed a spurt of replacement activities at the aspect of a replacement array of chemicals. Populations in industrialized countries stirred from rural areas to urban areas created larger amounts of waste and dioxide emissions. In Asia, as on date biomass burning and fuel burning are the foremost necessary sources of pollution that contribute to dioxide and GHGs. Copious of the heat radiation is absorbed by the water vapors whereas dioxide and completely different GHGs absorb the radiated energy not absorbed by water vapors. Biomass burning plays a significant role in volatilized pollution whereas fuel and biomass combustion contribute to particulate pollution (UNEP and C4, 2002).

Between 1971 and 2007, world emission doubled from developing countries, led by Asia, in line with exaggerated economic activity at a way faster rate. Between 1990 and 2007, dioxide emissions rose quite double for Asia because of rate of economic development notably among China and India (IEA, 2007). Fossil fuel production within the Middle East and Asia Pacific was increased throughout 2009, driven by growth in Iran, Qatar, India and China (EPA, 2010). New coal-fired power stations are being place in operation in China hebdomadally challenging of the ambitions of the United Nations Framework Convention on Climate Change (UNFCCC) and thus the urban Protocol challenge to merge desired economic development with environmental protection (UNEP, 2010). Asia has to boot some distinctive drawback within it relation to population distribution. Where, the South Asian Region is one amongst the foremost densely inhabited within the global. It possesses 3% of the world's landmass with 20% of world population. Population in cities of Asia and thus the Pacific region is higher than World Health Organization pointers. It absolutely was as high as 98% for India and 99% for China (UNEP and C4, 2002). Such population density is certainly vulnerable to pollution and global climate change.

Human activities finish in emissions of four principal GHGs: dioxide (CO_2), methane (CH_4), nitrous oxide (N_2O) and therefore the halocarbons (a cluster of gases fluorine, chlorine, and bromine). These gases accumulate among the atmosphere levels, inflicting concentrations to increase with time (Chandrappa et al., 2011).

- Carbon dioxide concentrations are increased because of fuel use, deforestation, and decay of organic matter.
- Methane concentrations are increased as a result of agriculture, fuel distribution, landfills, and anaerobic decomposition in wetlands.
- Nitrous chemical element compound is emitted by chemical/fertilizer manufacture and fuel burning. Natural processes like, volcanic

activity, fire make a reaction and unleash the nitrous oxide in the atmosphere and oceans.

- Halocarbon gas concentrations are increased primarily because of human activities.
- Main halocarbons embrace the chlorofluorocarbons (e.g., CFC-11 and CFC-12), that was used extensively as refrigeration agents and in several industrial processes.
- Gases are frequently made and destroyed within the atmosphere by series of chemical reactions within the layer of the Ozone, human activities have inflated in destroying Ozone by unleashing of gases like carbon monoxide, hydrocarbons, and nitrogen oxide that with chemically react with ozone.
- Water vapor is generated through chemical destruction of CH_4 among the layer, producing a touch amount of vapor.
- Aerosols are generated by every phylogeny activity (such as surface mining, fuel combustion, and industrial processes) and natural sources (mineral dirt free from the surface, ocean salt aerosols, biogenic emissions from the land and oceans; and dirt aerosols made by volcanic eruptions).

3.16 INDUSTRIAL REVOLUTION AND ASIA

The industrial revolution was introduced by Europeans into Asia among the last years of the 19th and thus the beginning of the twentieth century that saw the event of industries in India, China, and Japan. The achievable reasons for Asia not catching up with age in the 18th and early 19th century may are many countries were at a place in the management of colonial rule. The ulterior struggle of these countries for freedom what is more as being busy in the participation of wars rather than peace created Asian countries to catch-up industrial activity late.

The blessing in disguise to Asia in the twentieth century maybe a few economic homes wanted forward to supply polluting activities to avoid stringent environmental legislation and standards in their own countries. Completely different reasons embrace affordable labored probe for new market by recent corporations World Health Organization have achieved end of growth in their own countries as most of the folks in these countries possess materialistic luxury and thus does not turn out demand for cars, refrigerators, electronic merchandise, etc., among the 60s, concerning hour of the Chinese labor force was employed in agriculture, by 1990; the fraction of the labor force employed in agriculture had fallen to concerning time unit and by 2000 still further. The quick economic process is in the main because of the growth of the economic sector in absolute terms, of up to eight every year throughout the 70s. The economic process and thus the amplified integration among the global economy of various countries from Asia are tributary to the increase of international marine transport. Globally, big enterprises dominate industrial enterprise, but small- and medium-sized enterprises (SMEs) giant industries in developing countries. In India SMEs, have major shares in the metals, chemicals, and food industries (GOI, 2005).

China has 39.8 million SMEs accounting for 99% of the country's enterprises (APEC, 2002). One disadvantage with SMEs is lower adherence to environmental laws. Many SMEs are unorganized and can't register with government bodies to evade trade restrictions, paperwork, taxes, bribes, and work. Production of Natural Gas within the Middle East and the Asia Pacific exaggerated among the year 2009 because of growth in an Asian country, Qatar, India, and China (BP, 2010). China has become the world's largest energy-related GHG electrode surpassing the US. In 2007, China's energy sector accounted concerning 6.1 billion loads of carbon dioxide or concerning 20% initial of entire international energy-related dioxide emissions (WEO, 2009). China is presently the world's principal producer of cement, glass, steel, and ammonia (CSIS, 2010). If Asian countries still apace industrialize, transport demand will grow with extreme quickness over sequent several decades. It's apparently that combined radiate forcing from dioxide, CH_4 and N_2O concentration augmentation, are a minimum of five times quicker over the amount from 1960 to 1999 than over the last 40-year period for the length of the past two millennia before the industrial era (Jansen et al., 2007). Economic and increase in Asia over the last 30 years have been extraordinary. Whereas criterion economic indicators are growing constantly, indicators of resource and environmental qualities are decreasing, the reason typically remarked outdated technologies, varied unorganized small-scale units, and intuitional failure. Urbanized populations and poorly planned municipal development; 2.5 times augment within the use of traveler cars over the last two decades; haze pollution from forest fires in the Republic of Indonesia and other countries has resulted in urban pollution in Asia (Chandrappa et al., 2011).

3.17   URBANIZATION

As a result of urbanization, there is tremendous pressure on the surroundings because of the increase in vehicles and housing activities. The industrialization has to boot resulted in the formation of multiple nuclear families by ripping of joint family. Such development has created a demand for special vehicles and house domestic articles. Growing pressure to deliver merchandise and repair has to boot reduced time to cook and thus has increased quick nutrition business and restaurants and hotels in many cities. This means there is a ton of activity and combustion of fuel that has ultimately contributed to surroundings degradation. There is to boot drawback of the quality of constructions because of mushrooming flats. Exaggerated construction activities diminished the accessibility of quality sand around the cities. Construction corporations in such areas are combining soil with high fine particles to make concrete and mortar that ends in low bonding. The long-run impact may only be assessed throughout disasters once flats fall like a castle of playing cards. Completely different practices in construction like the use of freshwater for intermixing the separate material for construction will have an impact in the future because of the depletion of water resources. There is tremendous pressure on the hills because of mining and activity for supplying construction material (Chandrappa et al., 2011).

3.18   GLOBAL OIL

In 2009, international oil use declined by 1.2 million barrels per day (b/d), or 1.7%, however, world-refining capability grew by 2.2%, or 2 million b/d with the Asia-Pacific region accounting for quite 80% of the global growth, fundamentally because of an increase in India (BP, 2010). The software Industry boom was conjointly blessing in disguise for India that may supply the workforce at 6 to 10 times cheaper than the workforce cost in the USA. The fluency in English and talent to work exhausting with little further incentive has been the success story of computer software business in India. The event centers that employments round the clock typically demand a piece of over 12 hours from their staff. As a result, the operation population was able to generate a financial gain of 10–20 times that of their parents. This has exaggerated the shopping power of youngsters in a growing population of cities like Bangalore, Pune, Delhi, Noida, Gurgaon, Mumbai, and Hyderabad. The computer software giants have started development centers in less populates cities like Mysore, Indore, and Thiruvananthapuram to cut down costs.

From 1900 to 2000, world primary energy inflated quite denary, whereas the world population rose only fourfold from 1.6 to 6.1 billion (Sims et al., 2007). The perfect rate of growth among the last decade was in Asia. An oversized variety of the world's energy-intensive industries are presently been in developing countries with China being the world's largest producer of steel (IISI, 2005), aluminum, and cement (USGS, 2005).

3.19   CLIMATE CHANGE IS A MULTIPLIER OF HUMAN IMPACTS AND RISKS

Climate change is already seriously moving many folks nowadays and among the following 20 years, those affected will apparently quite double—making it the simplest rising humanitarian challenge of our time. Those seriously affected are in would really like of immediate help either following a weather-related disaster, or as a result of livelihoods are severely compromised by global climate change. The number of those severely affected by global climate change is over 10 times bigger than for example those injured in traffic accidents every year, and over the world-wide annual variety of recent malaria cases. Among the eventual 20 years, 1 in 10 of the world's present population may be directly and seriously affected. Already nowadays, several thousands of lives are lost once a year because of climate changes. This could rise to roughly 0.5 a million in 20 years. Over 9 in 10 deaths are related to gradual environmental degradation because of global climate change—in the main deficiency disease, diarrhea, malaria, with the remaining deaths being joined to weather-related disasters led to by global climate change. Economic losses are to global climate change presently amount to quite a 100 billion $ dollars every year that are quite the individual national GDPs of three-quarters of the world's countries.

This figure constitutes quite the complete of Official Development facilitate throughout a given year Already nowadays, over 0.5 a billion people are at extreme risk to the impacts of global climate change, and 6 in 10 people are vulnerable to global climate change throughout a physical and socio-economic sense. The bulk of the world's population doesn't have the capability to address the impact of global climate change while not suffering a doubtless irreversible loss of well-being or risk of loss of life. The populations most gravely in danger live in the poorest areas that are extremely at risk of global climate change specifically, the semi-arid solid ground belt countries from the Sahara to the Middle East and Central Asia, additionally as sub-Saharan Africa (SSA), South Asian, and tiny Island Developing States (Fust, 2009).

3.20 CLIMATE CHANGE DAMAGES HUMAN HABITAT

Increased temperatures manufacture rises in ocean level; soften glaciers, increase unpredictable weather events and alter rain patterns. They boot bring lots of frequent, lots of intense weather-related disasters. Most of the discovered increase in international average temperatures since the mid-20th century is improbably apparently joined to the rise in gas emission—emissions generated by human activities. These physical changes manifest themselves through gradual environmental degradation like activity and weather-related disasters like floods. Among the long run, potential big scale tipping-point events just like the quick melting of the Arctic and Greenland ice sheets, a retreat of the Amazon and thus the Boreal forests or a conclusion of the ocean current would each have a likely monumental impact on world climate. However, as these events are unlikely to occur among resultant 20 years (WFP, 2009).

Through a complicated set of effects, climate change impacts human health, livelihoods, safety, and society.

This chapter seeks to point the foremost reliable proof measure the human impact of events that will be attributed to climate change. Global climate change impacts on people the following ways that (WFP, 2009):

- **Food Security:** Lots of poor people, in particular kids, suffer from hunger because of reduced agricultural yield, mammal, and fish provide, as a result of environmental degradation.
- **Health:** Threats like diarrhea, malaria, respiratory disorder, and stroke have an impact on lots of people once temperatures rise.

- **Poverty:** Livelihoods are destroyed gain from agriculture, livestock, business enterprise, and fishing is lost because of weather-related disasters and activity.
- **Water:** Insufficiency of water results from a decline inside the general provider of freshwater and lots of frequent and severe floods and droughts.
- **Displacement:** Lots of climate-displaced people are expected because of water level rise activity and floods.
- **Security:** A lot of common people subsist under the continual threat of potential conflict because of migration, weather-related disasters, and water insufficiency.

3.21 CLIMATE CHANGE AND HEALTH PROBLEMS

Climate change is responsible for several 100 million more people stuffed with health problems and a number of 100,000 lives lost.

Every year, the health of 235 million people is likely be walking seriously full of gradual environmental degradation because of global climate change. This assumes that climate change affects malnutrition, diarrhea, and protozoa infection incidences. Moreover, among sequent year over 300,000 people are expected to die from health problems directly as a result of climate change. Malnutrition disease is the biggest burden in terms of deaths. Climate change is projected to cause over 150,000 deaths annually and nearly 45 million people are determinable to be malnourished as a result of global climate change, significantly because of reduced food provide and ablated gain from agriculture, mammal, and fisheries. Climate change-related diarrhea incidences are projected to amount to over 180 million cases annually, resulting in nearly 95,000 fatalities, notably because of sanitation issues joined to water quality and quantity. Climate change-triggered malaria infection outbreaks are determinable to own an impact on over 10 million people and kill roughly 55,000 (Fust, 2009).

3.22 FOOD SECURITY: CLIMATE CHANGE LEADS TO HUNGER

Weather-related disasters destroy crops and in the reduction of soil quality in a very variety of the world's poorest regions. Exaggerated temperatures, ablated rain, water shortages, and drought reduce yields and impact off mammal health. Climate activity fares away at the amount of arable land and thus the standard of the soil. Within the world's oceans, climate change and reef destruction cut back fish stock. The impacts are notably severe in developing regions like South Asia, SSA and thus the ground belt that stretches across the Sahara Desert and therefore the Middle East all the way to parts of China. Whereas hotter temperatures attracting attention in a lot of favorable agricultural conditions and exaggerated yield in some parts of North America and Russia, the global impact of climate change on overall food production is negative. The injury is incredibly severe among the world's poorest areas, where subsistence farmers get hit double by the less favorable growing conditions.

First, several might not have enough crop producers to feed their families. Second, the deficiency of their own crop may apparently force them to buy for food at a time once prices are high because of reduced international crop yields and increase. Over 900 million are chronically hungry today—many of them because of action. In 2008, the Food and Agriculture Organization (FAO) determinates that quite 900 million are afflicted with hunger, or concerning 13% of the global population. Of these tormented by hunger, 94% board developing nations. Most are subsistence farmers, landless or folks operating in fishery or forestry. The remainder lives in shanty cities on the fringes of urban areas. A quarter of the hungriest is kids. Global climate change is projected to be at the root of hunger and malnutrition disease for about 45 million folks, as a result of reduced agricultural yields of cereals, fruits, vegetables, livestock, and dairy, additionally the cash crops like cotton and fish which generate income. As an example, drought hurts crops in Africa where over 90% of farmers are small scale and concerning 65% of people's primary supply of financial gain is agriculture (IFPRI, 2004).

3.23 CLIMATE CHANGE AND THE CYCLE OF POVERTY

Because the poor tend to live in geographical and climatic regions that are naturally most prone to global climate change, their capability to adapt is certainly flooded by the impact of the changing conditions. They need the tiniest amount assets to trust among the event of a shock—whether or not it is a weather-related disaster, a nasty harvest, or a loved one falling unwell. These factors hinge on each other and build a perpetuating cycle of poorness that is robust to interrupt. Safety net structures like insurance are also for the foremost half unavailable to the world's poor. Many are subsistence farmers, fishermen, or have jobs among the business enterprise industry—vocations very captivated with natural resources just like the ocean, forests, and land for his or her livelihoods. Global climate change compounds existing poorness by destroying livelihoods. Specifically, rising temperatures, dynamical rain patterns, floods, droughts, and completely different weather-related disasters destroy crops and weaken or kill farm livestock. Rising temperatures and acidic oceans destroy coral reefs and accelerate the loss of fish stock. Loss of diversity, weather-related disasters like hurricanes and water level rise have durable negative impacts on business enterprise. The reef Alliance estimates that coral bleaching may result in billions of bucks in losses due to reduced multifariousness, coastal protection and income from reef fisheries and touristy. About 6–7 million losses are projected within the next 10 years if coral doesn't recover within the Philippines based on the net present value of the native diving business. Global climate change drives economic condition through a vicious circle of reduced crop yield and succeeding lower gain that leaves fewer resources for the next year's planting season. Concerning 60% exploit developing nations' personnel, about 1.5 billion people, are employed in agriculture, livestock, fisheries, and business enterprise. Most of the farmers carry unclean minimum production and losing a touch amount of their yield pushes them even further into poorness. Fishermen and other people employed in business enterprise lose gain or income concerning 60% exploit developing nations' personnel, about 1.5 billion people, is employed in agriculture, livestock, fisheries and business enterprise. Most of the farmers carry unclean minimum production and losing a touch amount of their yield pushes them even further into poorness. Fishermen and other people employed in business enterprise lose gain or income discharged.

The loss of diversity is worrisome not only because of its direct impacts on people's livelihoods, but to boot because of the intrinsic value of diversity and its crucial role in building the poor's resilience to action. Regeneration evolves as temperature change particularly alters ecosystems and reduces species diversity. As an example, species diversity assists in strengthening the ability for cod or lobster fishing resources to sustain stress and shocks. Theme self-regulating processes are crucial, just like the creation of natural carbon sinks that remove acid gas from the atmosphere. Exaggerated landscape diversity with varied plant species and natural lineation barriers like flowering tree forests can defend coastal inhabitants and their belongings from climate shocks like coastal storms and soil erosion. Having an assortment of ancient seeds to help confirm lots of drought resistant crop varieties is a lot of and a lot of very important to survival in drought-prone areas. There is nice cause for concern as a result of the IPCC estimates that 20–30 % exploit worldwide species are apparently to be at risk of extinction in this century (Foundation, 2009). Where, El Niño happens if ocean surface temperature rises by over 0.5°C across the central tropical Pacific Ocean. And some consultants counsel that El Niño frequency, length, and severity are increasing to warming. For that El Niño effects are usually stronger in South America than in North America, i.e., it's related to heat and extremely wet summers on the Peruvian and Ecuadorian coastline. It also has effects on world weather like making drier conditions in Northern Australia and wetter climate along the eastern African coastline.

3.24 HOW CLIMATE AFFECTS AGRICULTURE?

Climate change will have an impact on agriculture throughout a form of way in which. On the away side a precise vary of temperatures; warming tends to reduce yields because crops speed through their development, producing less grain within the operation. And better temperatures interfere with the power of plants to get and use moisture. Evaporation from the soil accelerates once temperatures rise and plants increase transpiration— that is, lose lots of moisture from their leaves. The combined impact is termed evapotranspiration, as world warming is probably going to extend rainfall, the net impact of upper temperatures on water accessibility may be a race between higher evapotranspiration and higher precipitation.

Typically, that race is won by higher evapotranspiration. However a key culprit in climate change—carbon emissions—can to boot facilitate agriculture by enhancing chemical process in many necessary, questionable C3, crops (such as wheat, rice, and soybeans). The science, however, is way from sure on the advantages of carbon fertilization. But we have a tendency to try to perceive that this development does not abundant facilitate C4 crops (such as sugarcane and maize), that account for concerning 25% of all crops by value (Cline, 2007).

3.25   FUTURE CLIMATE CHANGE COULD BE GREATER OR LESS THAN PRESENT-DAY BEST PROJECTIONS

Any action involves risk if its outcomes cannot be expected and thus the prospect of significant hurt cannot be dominated out. Uncertainty concerning the climate system does not decrease the risk associated with gas emissions, as a result of it works in every direction: climate change may persuade to be less severe than current estimates, but may addition convince be worse (Lewandowsky et al., 2014). While future changes from gas emissions are at the low end of the expected vary, a high-emissions pathway would still be enough to go to the global to temperatures it isn't seen for many ample years, well before humans evolved. Throughout this state of affairs, there is also no assurance that necessary hurt would not occur.

3.26   SCIENCE HAS AN IMPORTANT ROLE IN IDENTIFYING AND RESOLVING UNCERTAINTIES

Science contains a very important role in distinctive and determination uncertainties, and informing public policy on global climate change all societies routinely build picks to balance or minimize risk with only partial information of but these risks will play out (Murphy, 2012). Typically this can be often true in defense, finance, the economy and much of various areas. Societies have faced and created picks concerning mineral, lead, CFCs, and tobacco (Cagin, 1993; Gibson, 1997; Mincingvalley, 2002; Bellinger and Bellinger, 2006).

Although each case has distinctive aspects, all carried scientifically incontestable but hard-to-quantify risks, and were contentious, in common with global climate change. Mechanisms are placed in place nationwide and internationally to facilitate scientific input into higher cognition. Specially, the international Intergovernmental Panel on global climate change (IPCC) has prepared thorough, 'policy-neutral however policy-relevant' assessments of the state of knowledge and uncertainties of the science since 1990, with the foremost recent assessment completed in 2014. Australian scientists have created a significant contribution to the quality and integrity of these international IPCC assessments.

3.27 SOCIETIES FACE CHOICES ABOUT FUTURE CLIMATE CHANGE

Managing the risks from future human-induced global climate change will basically be supported some combination of four broad strategies:

- **Emissions Reduction:** Reducing global climate change by reducing gas emissions.
- **Sequestration:** Removing carbonic acid gas (CO_2) from the atmosphere into permanent natural science, biological or oceanic reservoirs.
- **Adaptation:** Responding to and addressing global climate change as a result of it happens, in either a planned or unplanned technique.
- **Solar Geo-Engineering:** Large-scale designed modifications to limit the amount of daylight reaching the planet, in a very shot to offset the results of current gas emissions (Lovelock, 2008; Honorary Society, 2009; IPCC, 2012).

Each embodies an oversized suite of specific decisions, with associated risks, costs, and edges. The four ways in which can have an impact on each other: as an example, doing nothing to cut back emissions would want exaggerated expenditure to adapt to global climate change, and exaggerated potentialities of the future resort to geo-engineering.

3.28 CHANGES ARE EVIDENT IN MANY PARTS OF THE CLIMATE SYSTEM

Changes in line with an increase in international temperature are discovered in many completely different components of the climate system:

- Mountain glaciers are shrinking and tributary to international low-lying rise since concerning 1850. Melting accelerated significantly among the 1990s (Cogley, 2009; Leclercq et al., 2011; Marzeion et al., 2012; Sarah Vaughan et al., 2013; Gardner et al., 2013; Hirabayashi et al., 2013).

- The Greenland and West Antarctic Sheets have every lost ice since 1990, a further tributary to low-lying rise. Typically this can be often from the exaggerated discharge of ice into the ocean, and to boot exaggerated surface melting in Greenland. The speed of loss from Greenland looks to be increasing (Shepherd et al., 2012; Sarah Vaughan et al., 2013).

- The area of the ocean lined by ocean ice has ablated significantly since 1987 throughout the year and notably in summer (Stroeve et al., 2007; Comiso et al., 2008).

- The thickness of the ice has addition ablated by over 30% over the last 30 years (Kwok, 2007; Rothrock et al., 2008; Comiso, 2011; Wadhams et al., 2011).

- Within the Southern Ocean, there are durable regional variations among the changes to areas lined by ocean ice (Stammer et al., 2012), but a few increases in total coverage (Comiso and Nishio, 2008), driven by shifts in winds and ocean currents throughout warming the Southern Ocean. Strengthening circumpolar winds around Antarctica has addition been joined partially to dilution of the ozonosphere layer.

- The amount of vapor among the atmosphere has exaggerated since the 1980s (Sherwood et al., 2010; Chung et al., 2014), that's in line with the hotter air.

- The surface of the ocean in rainy elements of the earth is turning into less salty, that's in line with recent dilution from exaggerated rain (Durack et al., 2012).

- Some ocean currents have changed in response to changes in surface winds, ocean temperature, and ocean saltiness. The changes embrace a southward shift of the Antarctic Circumpolar Current (Böning et al., 2008; Sokolov and Rintoul, 2009) and increasing southward penetration of the East Australian Current (Ridgway, 2007).

- An increasing type of plants and animals, on land and among the oceans, are undergoing shifts in their distribution and lifecycles that are in line with discovered temperature changes (Poloczanska et al., 2013; Van der Wal et al., 2013).

3.29 HAS CLIMATE WARMING RECENTLY STOPPED?

According to most estimates (Grant and Stefan, 2011; Morice et al., 2012; NASA-GISS, 2013; NOAA-NCDC, 2013; Kosaka and Xie, 2013; Cowtan, 2014), the speed of average surface warming has slowed since 2001; despite current rises in greenhouse gases. This lag is in line with known climate variability. Indeed, decades of little or no temperature trend is also seen throughout the last century, superimposed on the long-term warming trend (Easterling and Wehner, 2009).

Two main factors have contributed to the slowed surface warming: initial, decadal variability in the ocean and the atmosphere system redistributed heat within the ocean, particularly within the eastern and central Pacific (Kosaka and Xie, 2013; Trenberth and Fasullo, 2013; European Nation et al., 2014). This has caused warming at depth and cooling of surface waters and therefore the lower atmosphere during this region. Second, many temporary world cooling influences have acquired play together with remarkably weak solar activity, inflated aerosol production, and volcanic activity, exaggerated aerosol production, and volcanic activity (Solomon et al., 2010, 2011; Helmut Schmidt et al., 2014; Santer et al., 2014). None of those influences is maybe going continue over futurity. Moreover, despite the lag in warming at the surface, there are continued can increase in heat extremes (Seneviratne et al., 2014) and among the physical property of the oceans (Lyman et al., 2010; Purkey et al., 2010; Meehl et al., 2011; Levitus et al., 2012; Watanabe et al., 2013; Balmaseda et al., 2013; England et al., 2014), what is more, as rising ocean levels, shrinking Arctic sea-ice, and current soften of ice sheets and glaciers. Some models predict that, once this lag ends, revived warming is quick (England et al., 2014).

3.30 EVIDENCE OF CLIMATE CHANGE

The proof of climate change extends well on the way aspect can increase in world surface temperatures; it is included (EPA, 2010):

- Changing precipitation patterns.
- Melting ice in the Arctic;
- Melting glaciers around the world;

- Increasing ocean temperatures;
- Rising water level around the world;
- Changing the action of the oceans because of elevated carbonic gas among the atmosphere;
- Responses by plants and animals as shifting ranges.

3.31   PROJECTIONS OF CLIMATE CHANGE

At this rate, the Earth's world average temperature is projected to rise from three to seven °F by 2100, and it will get even hotter then. As a result of the climate, change continues in heating, lots of changes are expected to occur, and much of the effects will become lots of pronounced over time. As an example, heat waves are expected to become lots of common, severe, and longer-lasting. Some storms are apparently to become stronger and lots of frequent, increasing the chances of flooding and injury in coastal communities (EPA, 2010).

Global climate change will have an impact on altogether completely different regions, ecosystems, and sectors of the economy in some ways in which, relying not only on the sensitivity of those systems to action, but to boot on their ability to adapt to risks and dynamical conditions. Throughout history, societies, and ecosystems alike have shown exceptional capability to reply to risks and adapt to altogether different climates and environmental changes. Today, the effects of global climate change have already been discovered, and thus the speed of warming has exaggerated in recent decades. For this reason, human-caused global climate change represents a major challenge—one that may want new approaches and ways in which of thinking to create certain the continued health, welfare, and productivity of society and thus the natural surroundings.

3.32   CONCLUSION

Climate change refers to significant, long-term modifications in the global weather. The worldwide weather is the connected system of solar, earth and oceans, wind, rain and snow, forests, deserts and savannas, and the entirety people do, too. The climate of a place, can be described as its rainfall, changing temperatures in the course of the year and other. But the international weather is greater than the "common" of the climates of particular places. In which a description of the international climate consists for example from the rising temperature of the Pacific which feeds typhoons which blow harder, drop extra rain and cause extra damage, however also shifts worldwide ocean currents that soften Antarctica ice which slowly makes sea level rise until to some cities will be below water. It is this systemic connectedness that makes worldwide climate alternate so crucial and so complicated.

Global warming is the slow growth inside the common temperature of the earth's surroundings because an increased quantity of the energy (heat) striking the earth from the solar is being trapped inside the environment and no longer radiated out into space. The earth's ecosystem has constantly acted like a greenhouse to capture the sun's heat, ensuring that the earth has enjoyed temperatures that approved the emergence of existence forms as we know them, consisting of humans. Without our atmospheric greenhouse the earth would be very cold. Global warming, however, is the equal of a greenhouse with high performance reflective glass mounted the wrong way around. The best evidence of this will come from a terrible cooling occasion that came about some 1,500 years ago. Two big volcanic eruptions, 365 days after another positioned a lot black dust into the upper atmosphere that little sunlight could penetrate. Temperatures plummeted. Crops failed. People died of starvation and the Black Death began its march. As the dirt slowly fell to earth, the sun become again able to warn the arena and lifestyles returned to normal. Today, we have the alternative problem. Today, the problem is not that too little sun warmth is reaching the earth, however that an excessive amount of is being trapped in our ecosystem. So much warmness is being kept inside greenhouse earth that the temperature of the earth goes up faster than at any preceding time in history.

Heat is energy and when you add power to any system changes occur. Because all systems in the global climate system are connected, adding heat energy causes the global climate as an entire to exchange. Much of the world is blanketed with ocean which heats up. When the ocean heats up, more water evaporates into clouds. Where storms like hurricanes and typhoons are forming, the end result is more strength-extensive storms. A warmer ecosystem makes glaciers and mountain snow packs, the Polar ice cap, and the excellent ice protect jutting off of Antarctica melt elevating sea levels. Changes in temperature change the excellent patterns of wind that carry the monsoons in Asia and rain and snow around the arena, making drought and unpredictable weather more common. This is why scientists have stopped focusing simply on global warming and now attention on the bigger subject matter of weather exchange. There are three positions on international warming: (1) that international warming is not occurring and so neither is climate exchange; (2) that worldwide warming and climate exchange are going on, however those are natural, cyclic activities unrelated to human interest; and (3) that global warming is happening as a result more often than not of human activity and so climate change is likewise the end result of human activity.

They declare that nothing is going on is very hard to defend inside the face or loads of visual, land-based totally and satellite facts that simply suggests rising average sea and land temperatures and shrinking ice masses. They declare that the determined worldwide warming is natural or at the least not the result of human carbon emissions (see Climate Skeptics below) focuses on facts that suggests that world temperatures and atmospheric CO_2 levels were equally high or higher in the past. They additionally factor to the properly understood consequences of solar on the amount of radiation striking the earth and the fact that these days the sun has been especially active.

In general, climate scientists and environmentalists either (1) dispute the statistics based totally on, for example, new ice core records or (2) advocate that the timing issue – that is, the rapidity with which the globe has warmed and the climate changed truly do not suit the model of preceding natural occasions. They observe also that compared to different stars the sun is virtually very stable, various in energy output by just 0.1% and over a relatively brief cycle of 11 to 50 years quite unrelated to worldwide warming as a whole. The records strongly show that solar activity impacts the global weather in many important ways, however isn't always a factor within the systemic change over time that we call global warming.

KEYWORDS

- carbon dioxide (CO_2)
- carbon emissions
- fossil fuels
- global climate change
- greenhouse-gas (GHG)
- industrial revolution

REFERENCES

Anderegg, W. R. L., Prall, J. W., Harold, J., & Schneider, S. H., (2010). Expert credibility in climate change. *Proceedings of the National Academy of Sciences, 107*, 12107–12109.

Andrew, H., (2015). *The Science of Climate Change, Questions and Answers*. Australian Academy of Science, ISBN: 978-0-85847-413-0. www.science.org.au/climatechange (accessed on 4 January 2020).

APEC, (2002). *Profiles in SMEs and SME Issues, 1990–2000*. Asia-Pacific Economic Cooperation, Singapore, World Scientific Publishing.

Balmaseda, M. A., Trenberth, K. E., & Kallen, E., (2013). Distinctive climate signals in reanalysis of global ocean heat content. *Geophysical Research Letters, 40*, 1754–1759.

Bellinger, D. C., & Bellinger, A. M., (2006). Childhood lead poisoning: The torturous path from science to policy. *The Journal of Clinical Investigation, 116*, 853–857. doi: 10.1172/JCI28232.

Böning, C. W., Dispert, A., Visbeck, M., Rintoul, S. R., & Schwarzkopf, F. U., (2008). The response of the Antarctic circumpolar current to recent climate change. *Nature Geosci., 1*, 864–869, doi: 10.1038/ngeo362.

British Petroleum (BP), (2010). *BP Statistical Review of World Energy*. BP Oil Company Ltd., London.

Cagin, S., & Dray, P., (1993). *Between Earth and Sky: How CFCs Changed our World and Endangered the Ozone Layer.* Pantheon.

Chandrappa, R., Gupta, S., & Kulshrestha, U. C., (2011). *Coping with Climate Change*. Springer-Verlag Berlin Heidelberg. doi: 10.1007/978-3-642-19674-4_2.

Chung, E. S., Soden, B., Sohn, B. J., & Shi, L., (2014). Upper tropospheric moistening in response to anthropogenic warming. *Proceedings of the National Academy of Sciences, 11*, 11636–11641, doi: 10.1073/pnas.1409659111.

Cline, W. R., (2007). *Global Warming and Agriculture: Impact Estimates by Country*. Washington: Center for Global Development and Peterson Institute for International Economics.

Cogley, J. G., (2009). Geodetic and direct mass-balance measurements: Comparison and joint analysis. *Annals of Glaciology, 50*, 96–100. doi: 10.3189/172756409787769744.

Comiso, J. C., & Nishio, F., (2008). Trends in the sea ice cover using enhanced and compatible AMSR-E, SSM/I, and SMMR data. *Journal of Geophysical Research: Oceans, 113*, C02S07, doi: 10.1029/2007JC004257.

Comiso, J. C., (2011). Large decadal decline of the Arctic multiyear ice cover. *Journal of Climate, 25*, 1176–1193. doi: 10.1175/JCLI-D-11-00113.1.

Comiso, J. C., Parkinson, C. L., Gersten, R., & Stock, L., (2008). Accelerated decline in the Arctic sea ice cover. *Geophysical Research Letters, 35*, L01703. doi: 10.1029/2007GL031972.

Cook, J., et al., (2013). Quantifying the consensus on anthropogenic global warming in the scientific literature. *Environmental Research Letters, 8*, 024024.

Cowtan, K., & Way, R. G., (2014). Coverage bias in the HadCRUT4 temperature series and its impact on recent temperature trends. *Quarterly Journal of the Royal Meteorological Society.* doi: 10.1002/qj.2297.

CSIS, (2010). *Centre for Strategic International Studies.* Asia's Response to Climate Change and Natural Disasters, Implications for an Evolving Regional Architecture.

Doran, P. T., & Zimmerman, M. K., (2009). Examining the scientific consensus on climate change. *Eos, Transactions American Geophysical Union, 90*, 22–23. doi: 10.1029/2009EO030002.

Durack, P. J., Wijffels, S. E., & Matear, R. J., (2012). Ocean salinities reveal strong global water cycle intensification during 1950 to 2000. *Science, 336*, 455–458.

Easterling, D. R., & Wehner, M. F., (2009). Is the climate warming or cooling? *Geophysical Research Letters, 36*. doi: 10.1029/2009GL037810.

EEA, (2015). *Atmospheric Greenhouse Gas Concentrations.* European Environment Agency, EEA, Copenhagen.

England, M. H., et al., (2014). Recent intensification of wind-driven circulation in the Pacific and the ongoing warming hiatus. *Nature Clime. Change, 4*, 222–227. doi: 10.1038/nclimate2106.

EPA, (2010). *Climate Change Science Facts.* Environmental Protection Agency, www.epa.gov/climatechange (accessed on 16 January 2020).

FAO, (2007). *The United Nations, Adaptation to Climate Change in Agriculture, Forestry, and Fisheries: Perspective, Framework, and Priorities*, Viale Delle Terme di Caracalla-00100 Rome, Italy.

Foundation, (2009). *Impacts: Extreme Weather the El Nio Connection.* David Suzuki Foundation. http://www.davidsuzuki.org/Climate_Change/Impacts/Extreme_Weather/El_Nino.asp (accessed on 16 January 2020).

Gardner, A. S., et al., (2013). A reconciled estimate of glacier contributions to sea level rise: 2003 to 2009. *Science, 340*, 852–857. doi: 10.1126/science.1234532.

Gibson, B., (1997). An introduction to the controversy over tobacco. *Journal of Social, Issues, 53*, 3–11. doi: 10.1111/j.1540-4560.1997.tb02428.x.

GOI, (2005). *Annual Report, 2004–2005 of the Ministry of Environment and Forest.* Government of India, New Delhi.

Grant, F., & Stefan, R., (2011). Global temperature evolution 1979–2010. *Environmental Research Letters, 6*, 044022.

Hamada, Y. M. (2018). Special pictures album, Took by Youssef M. Hamada, Egypt, 2018.

Hamilton, K., (1998). *The Oil Industry and Climate Change.* A Greenpeace briefing, Greenpeace International. www.greenpeace.org (accessed on 4 January 2020).

Hirabayashi, Y., Zhang, Y., Watanabe, S., Koirala, S., & Kanae, S., (2013). Projection of glacier mass changes under a high emission climate scenario using the global glacier model HYOGA2. *Hydrol. Res. Lett., 7*, 6–11.

Holli, R., (2007). *Earth Observatory's Global Warming*. EOS project science office located at NASA Goddard Space Flight Center. http://earthobservatory.nasa.gov/Features/GlobalWarming (accessed on 4 January 2020).

IEA, (2007). World *Energy Outlook 2007 Edition, China and India Insights* (p. 600). International Energy Agency (IEA), Head of Communication and Information Office, 9 rue de la Federation, 75739 Paris Codex 15, France. ISBN: 9789264027.

IFPRI, (2004). *Ending Hunger in Africa Prospects for the Small Farmer.* International Food Policy Research Institute. http://www.ifpri.org/pubs/ib/ib16.pdf (accessed on 16 January 2020).

IISI, (2005). *World Steel in Figures, 2005: International Iron and Steel Institute (IISI).* Brussels. Intergovernmental Panel on Climate Change (IPCC) – 2007a: Climate change synthesis report – 2007.

IPCC, (2007). Summary for policymakers. In: Metz, B., Davidson, O. R., Bosch, P. R., Dave, R., & Meyer, L. A., (eds.), *Climate Change 2007: Mitigation. Contribution of Working Group III to the Fourth Assessment Report of the Intergovernmental Panel on Climate Change.* Cambridge University Press, Cambridge, United Kingdom and New York, NY, USA.

IPCC, (2012). *IPCC Expert Meeting on Geo-Engineering (Lima, Peru, 20–22 June 2011): Meeting Report.* Potsdam Institute for Climate Impact Research, Potsdam, Germany.

Jansen, E., Overpeck, J., Briffa, K. R., Duplessy, J. C., Joos, F., Masson-Delmotte, V., et al., (2007). Palaeo-climate. In: Solomon, S., Qin, D., Manning, M., Chen, Z., Marquis, M., Averyt, K. B., Tignor, M., & Miller, H. L., (eds.), *Climate Change 2007: The Physical Science Basis. Contribution of Working Group I to the Fourth Assessment Report of the Intergovernmental Panel on Climate Change.* Cambridge University Press, Cambridge, United Kingdom, and New York, USA.

Klaus, T., (2003). *UNEP, Information Unit for Conventions.* United Nations Environment, Program, CH-1219 Geneva, Switzerland. Web: www.unep.ch/conventions (accessed on 16 January 2020).

Kosaka, Y., & Xie, S. P., (2013). Recent global-warming hiatus tied to equatorial Pacific surface cooling. *Nature, 501*, 403–407. doi: 10.1038/nature12534.

Kwok, R., (2007). Near zero replenishment of the Arctic multiyear sea ice cover at the end of 2005 summer. *Geophysical Research Letters, 34*, L05501. doi: 10.1029/2006GL028737.

Lars Nordberg, (2010). Air Pollution Promoting Regional Cooperation, United Nations Environment Programme, ISBN: 978-92-807-3093-7, 2010.

Leclercq, P. W., Oerlemans, J., & Cogley, J. G., (2011). Estimating the glacier contribution to sea-level rise for the period 1800–2005. *Surveys in Geophysics, 32*, 519–535. doi: 10.1007/s10712–011–9121–7.

Lepetz, V., Massot, M., Schmeller, D. S., & Clobert, J., (2009). Biodiversity monitoring: Some proposals to adequately study species' responses to climate change. *Biodiversity and Conservation, 18*, 3185–3203.

Levitus, S., et al., (2012). World ocean heat content and thermo steric sea-level change (0–2000 m), 1955–2010. *Geophysical Research Letters, 39*, L10603. doi: 10.1029/2012GL051106.

Lewandowsky, S., Risbey, J., Smithson, M., Newell, B., & Hunter, J., (2014). Scientific uncertainty and climate change: Part I. Uncertainty and unabated emissions. *Climatic Change, 124*, 21–37. doi: 10.1007/s10584–014–1082–7.

Lovelock, J. E., (2008). A geo-physiologist's thoughts on geo-engineering. Philosophical transactions of the Royal Society A. *Mathematical Physical and Engineering Sciences, 366*, 3883–3890.

Lyman, J. M., et al., (2010). Robust warming of the global upper ocean. *Nature, 465*, 334–337. doi: 10.1038/nature09043.

Marzeion, B., Jarosch A. H., & Hofer, M., (2012). Past and future sea-level change from the surface mass balance of glaciers. *The Cry Sphere, 6*, 1295–1322. doi: 10.5194/tc-6-1295–2012.

Meehl, G. A., Arblaster, J. M., Fasullo, J. T., Hu, A., & Trenberth, K. E., (2011). Model-based evidence of deep-ocean heat uptake during surface-temperature hiatus periods. *Nature Clim. Change, 1*, 360–364. doi: 10.1038/nclimate1229.

Mimura, N., (2010). Scope and roles of adaptation to climate change. In: Sumi, A., Fukushi, K., & Hiramatsu, A., (eds.), *Adaptation and Mitigation Strategies for Climate Change.* Springer Tokyo Berlin Heidelberg New York, Ch. 9.

Montoya, J. M., Emmerson, M. C., Sole, R. V., & Woodward, G., (2005). Perturbations and indirect effects in complex food webs. In: De Ruiter, P., Wolters, V., & Moore, J. C., (eds.), *Dynamic Food Webs: Multispecies Assemblages, Ecosystem Development, and Environmental Change.* New York, NY: Academic Press.

Morice, C. P., Kennedy, J. J., Rayner, N. A., & Jones, P. D., (2012). Quantifying uncertainties in global and regional temperature change using an ensemble of observational estimates: The HadCRUT4 data set. *Journal of Geophysical Research-Atmospheres, 117*, doi: 10.1029/2011jd017187.

Nakicenovic, N. (2007). World Energy Outlook 2007: China and India Insights. Paris: IEA/OECD. ISBN 978-92-64-02730-5.

NASA-GISS, (2013). *GISS Surface Temperature Analysis, (GISTEMP).* Goddard Institute for Space Studies, National Aeronautics and Space Administration, New York City, NY, USA. http://data.giss.nasa.gov/gistemp/ (accessed on 16 January 2020).

NOAA-NCDC, (2013). *Global Surface Temperature Anomalies.* National Climatic Data Center, National Oceanic and Atmospheric Administration, USA, Boulder, CO, USA. http://www.ncdc.noaa.gov/cmb-faq/anomalies.php (accessed on 16 January 2020).

Poloczanska, E. S., et al., (2013). Global imprint of climate change on marine life. *Nature Clime, Change, 3*, 919–925. doi: 10.1038/nclimate1958.

Purkey, S. G., & Johnson, G. C., (2010). Warming of global abyssal and deep southern ocean waters between the 1990s and 2000s: Contributions to global heat and sea-level rise budgets. *Journal of Climate, 23*, 6336–6351. doi: 10.1175/2010JCLI3682.1.

Raymond, M., (2012). Managing risk under uncertainty. In: Measham, T., & Lockie, S., (eds.), *Risk and Social Theory in Environmental Management* (pp. 17–26). (CSIRO Publishing, 2012).

Rebecca, J., M., (2010). Anthropogenic impacts on tropical forest biodiversity: A network structure and ecosystem functioning perspective. *Philosophical Transactions of the Royal Society B, 365*, 3709–3718.

Ridgway, K. R., (2007). Long-term trend and decadal variability of the southward penetration of the East Australian Current. *Geophysical Research Letters, 34*, L13613. doi: 10.1029/2007GL030393.

Rothrock, D. A., Percival, D. B., & Wensnahan, M., (2008). The decline in Arctic sea-ice thickness: Separating the spatial, annual, and inter-annual variability in a quarter-century of submarine data. *Journal of Geophysical Research: Oceans, 113*, C05003. doi: 10.1029/2007JC004252.

Rouven, D., Daniel, H., & Svein, L., (2014). Attitudes, efficacy beliefs, and willingness to pay for environmental protection when traveling, downloaded from thr.sagepub.com at the University of Bergen on September 18, 2015. *Tourism and Hospitality Research, 15*(4), 281–292.

Royal Society, (2009). *Geo-Engineering the Climate: Science, Governance, and Uncertainty* (pp. 12–82). Royal Society, London.

Santer, B. D., et al., (2014). Volcanic contribution to decadal changes in tropospheric temperature. *Nature Geosci., 7*, 185–189. doi: 10.1038/ngeo2098.

Schmidt, G. A., Shindell, D. T., & Tsigaridis, K., (2014). Reconciling warming trends. *Nature Geosci., 7*, 158–160. doi: 10.1038/ngeo2105.

Seneviratne, S. I., Donat, M. G., Mueller, B., & Alexander, L. V., (2014). No pause in the increase of hot temperature extremes. *Nature Climate Change, 4*, 161–163.

Shepherd, A., et al., (2012). A reconciled estimate of ice-sheet mass balance. *Science, 338*, 1183–1189. doi: 10.1126/science.1228102.

Sherwood, S. C., Roca, R., Weckwerth, T. M., & Andronova, N. G., (2010). Tropospheric water vapor, convection, and climate. *Reviews of Geophysics, 48*, RG2001. doi: 10.1029/2009RG000301.

Sims, R. E. H., Schock, R. N., Adegbululgbe, A., Fenhann, J., Konstantinaviciute, I., Moomaw, W., et al., (2007). Energy supply. In: Metz, B., Davidson, O. R., Bosch, P. R., Dave, R., & Meyer, L. A., (eds.), *Climate Change 2007: Mitigation: Contribution of Working Group III to the Fourth Assessment Report of the Intergovernmental Panel on Climate Change*. Cambridge University Press, Cambridge, United Kingdom and New York, NY, USA.

Sokolov, S., & Rintoul, S. R., (2009). Circumpolar structure and distribution of the Antarctic circumpolar current fronts: Mean circumpolar paths. *Journal of Geophysical Research: Oceans, 114*, C11018. doi: 10.1029/2008JC005108.

Solomon, S., et al., (2010). Contributions of stratospheric water vapor to decadal changes in the rate of global warming. *Science, 327*, 1219–1223.

Solomon, S., et al., (2011). The persistently variable background stratospheric aerosol layer and global climate change. *Science, 333*, 866–870.

Stammerjohn, S., Massom, R., Rind, D., & Martinson, D., (2012). Regions of rapid sea ice change: An inter-hemispheric seasonal comparison. *Geophysical Research Letters, 39*, L06501. doi: 10.1029/2012GL050874.

Stevens, B., & Bony, S., (2013). What are climate models missing? *Science, 340*, 1053–1054.

Stroeve, J., Holland, M. M., Meier, W., Scambos, T., & Serreze, M., (2007). Arctic sea ice decline: Faster than forecast. *Geophysical Research Letters, 34*: L09501. doi: 10.1029/2007GL029703.

Trenberth, K. E., & Fasullo, J. T., (2013). An apparent hiatus in global warming? *Earth's Future, 1,* 19–32. doi: 10.1002/2013EF000165.

TRS, (2014). *The Royal Society.* A short guide to climate science. www.royalsociety.org (accessed on 4 January 2020).

Twee, D. G., (2002). Asbestos and its lethal legacy. *Nat. Rev. Cancer, 2,* 311–314.

UNEP and C4, (2002). *The Asian Brown Cloud: Climate and Other Environmental Impacts UNEP.* Nairobi.

UNEP, (2010). *Air Pollution Promoting Regional Cooperation.*

USGS, (2005). *Minerals Yearbook: 2004.* US Geological Survey Reston, VA, USA. http:// minerals.usgs.gov/minerals/pubs/myb.html (accessed on 4 January 2020).

Van Der Wal, J., et al., (2013). Focus on poleward shifts in species' distribution underestimates the fingerprint of climate change. *Nature Clim. Change, 3,* 239–243. doi: 10.1038/nclimate1688.

Vaughan, D. G., Comiso, J. C., Allison, I., Carrasco, J., Kaser, G., Kwok, R., et al., (2013). Observations: Cry sphere. In: Stocker, T. F., Qin, D., Plattner, G. K., Tignor, M., Allen, S. K., Boschung, J., et al., (eds.), *Climate Change 2013: The Physical Science Basis: Contribution of Working Group I to the Fifth Assessment Report of the Intergovernmental Panel on Climate Change* (Ch. 4, pp. 317–382). Cambridge University Press.

Wadhams, P., Hughes, N., & Rodrigues, J., (2011). Arctic sea ice thickness characteristics in winter 2004 and 2007 from submarine sonar transects. *Journal of Geophysical Research: Oceans, 116,* C00E02. doi: 10.1029/2011JC006982.

Walter Fust, (2009). The anatomy of a silent crisis. *Human Impact Report: Climate Change.* The anatomy of a silent crisis Published by the Global Humanitarian Forum—Geneva-©2009. www.ghf-ge.org (accessed on 4 January 2020).

Watanabe, M., et al., (2013). Strengthening of ocean heat uptake efficiency associated with the recent climate hiatus. *Geophysical Research Letters, 40,* 3175–3179. doi: 10.1002/ grl.50541.

WEO, (2009). World energy outlook. *Annex A: Tables for Reference Scenario Projections,* p. 647.

WFP, (2009). *Who Are the Hungry?.* World Food Program. http://www.wfp.org/hunger/ who-are (accessed on 16 January 2020).

Wikipedia, (2012). http://en.wikipedia.org/wiki/Albedo (accessed on 4 January 2020).

Yodzis, P., (2000). Diffuse effects in food webs. *Ecology, 81,* 261–266.

CHAPTER 4

Climatic Impacts on Global Warming

ABSTRACT

Global warming could also be a gradual increase at the earth's surface temperature. It's believed to result in a piece from a build-up of heat-trapping greenhouse gases (such as carbonic acid gas and gas oxide) emitted by human activities likewise as fuel burning and land clearing (IPCC, 2007). Continued heating has many damaging effects—it disrupts the ecosystems that provide us with water and food and exacerbates the prevailing environmental stresses like geological process, declining water quality, and pollution. The prevalence of some malady and different threats to human health are to some extent to warming.

Developed countries have out the largest acid gas emissions; but, regions most severely affected are those who not emit the greenhouse gases. Africa is emitting low greenhouse gases but hit hardest by warming. Thus, developed nations should not exclusively bear responsibility for inflicting heating but to boot fight against it. To tackle heating, the Kyoto Protocol (KP) set target levels for industrial nations to reduce gas emissions; however, the achievements created for the current point, significantly with regard to lots of simply distribution of contemporary Development Mechanism comes and don't seem to be best.

4.1 BACKGROUND

Global Warming is caused as a result of the rise of the everyday temperature on Earth. As a result of this, the Earth is getting hotter and disasters like hurricanes, droughts, and floods are getting further frequent. Over the last 100 years, the everyday air temperature near the Earth's surface has up by slightly, however, 1° or 1.3° Fahrenheit.

Heating is that the cause, world climate change is the result. Scientists sometimes choose to speak about international climate change instead of heating, as a result of higher international temperatures don't primarily mean that it will be hotter at any given time at every location on Earth. Warming is strongest at the Earth's Poles. In recent years, fall air temperatures are at a record 9°F (5°C) on top of traditional at intervals the Arctic, in step with the U.S. National Oceanic and atmospherically Administration. But changing wind patterns would possibly mean that the warming Arctic, for example, winds up in colder winters in continental Europe. Regional climates will modification nevertheless, but in really alternative ways in which. Some regions like regions of Northern Europe or West Africa will possibly get wetter, whereas completely different regions similar to the Mediterranean or Central Africa will presumably receive less rainfall. Melting ice is the foremost visible impact of a warming climate. The international organization Panel on international climate change finds that average Arctic temperatures have magnified at nearly double the world average rate in the past 100 years (Venkataramanan and Smitha, 2011). Figure 4.1 shows the Nile at sunset in Upper Egypt, Egypt.

FIGURE 4.1   The Nile at sunset, Upper Egypt, Egypt (Hamada, 2018).

4.2 SCOPE OF THE CHAPTER

The revolution of the twentieth century has allowed the farmers of the earth to use chemical fertilizers and machines to provide far more food than they ever did before. One in each of the primary components of the revolution has been the event of part fertilizers that dramatically accelerate the growth and productivity of plants in the field. Plants fix, or capture, part on their own additionally, but revolution technologies became therefore common that humans are presently adding plenty of part to the earth than all of the plants at intervals the world combined (Venkataramanan and Smitha, 2011).

Climate change incorporates an enormous impact on agricultural productivity and thence shortages of food provide a rise in food imports, which in turn causes an increase in food costs. Yields in many countries of Africa are reduced because of a lack of rain, hotter temperature, and an increase in evaporation. The United Nations (UN) expected that cereal crop yields would decline by up to 5% by the 2080s with subsistence crops like sorghum in Sudan, Ethiopia, Eritrea, and Zambia; maize in Ghana; millet in Sudan and groundnuts at Gambia (Elasha et al., 2006). In some African countries, yields from rain-fed agriculture can be reduced by up to 500% by 2020 (IPCC, 2007).

The implications of food insecurity in Africa are evident within intervals the incidence of hunger and malnourishment. In social insurance administration—the region is the simplest prevalence of malnourishment—1 in 3 is individuals in need of access to decent food (FAO, 2006).

4.3 A PRIMER ON CLIMATE CHANGE

The global climate change would increase the amount of people laid low with death, malady, and injury from heat waves, floods, storms, and droughts Floods are low-possibility and high-impact proceedings, which can overcome physical infrastructure and human communities. Major storm and flood disasters have occurred at intervals for the last 20 years. Vulnerability to weather disasters depends on the attributes of the person in peril, likewise as where they live and their age, additionally as totally different social and environmental factors. High-density populations in low-lying coastal regions experience a high health burden from weather disasters. Hot days, hot nights, and heat waves became lots of frequent. Heat waves are associated with mark short can increase in mortality. In some regions, changes in temperature and precipitation are projected to increase the frequency and severity of facet events. Forest and bush fires cause burns, suffering, from smoke inhalation and totally different injuries. Background levels of ground-level gas have up since pre-industrial times as results of increasing emissions of carbonic acid gas and element oxides. This trend is anticipated to continue into the mid-21st century (Venkataramanan and Smitha, 2011).

4.4   GREEN HOUSE EFFECT

Once daylight reaches the surface of land some of it absorbed and warms land, most of the rest is radiated back to the atmosphere at an extended wavelength than the sun light-weight and a variety of those longer wavelengths are absorbed by greenhouse gases at intervals the atmosphere before they are lost to space. The absorption of this radio radiation energy warms the atmosphere. These greenhouse gases act kind of a mirror and replicate back to the world variety of the heat energy which could preferably be lost up to space. The reflective back of heat energy by the atmosphere is termed the greenhouse effect. The most natural greenhouse gases are live vapors, which causes regarding 36–70% of the physical phenomenon on Earth (not likewise as clouds); gas greenhouse gas that causes 9–26%; carbonic acid gas, that causes 4–9%, and ozone, which causes 3–7%. It's infeasible to state that a certain gas causes a certain proportion of the atmospherical phenomenon, as results of the influences of the various gases are not additive. Totally different greenhouse gases embrace, but are not restricted to, inhalation general an aesthetic, fluoride, hydrofluorocarbons, perfluorocarbons and chlorofluorocarbons (CFCs). Just about 100% of the determination temperature increase over the last 50 years has been as a result of the increase at the greenhouse gas concentrations like vapors, carbonic acid gas. The greenhouse is those gases that contribute to the atmospheric phenomenon, the largest offer of greenhouse gas is the burning of fossil fuels leading to the emission of gas (Venkataramanan and Smitha, 2011).

4.4.1   CAUSES OF WORLD WARMING

The buildup of greenhouse gas at the atmosphere, primarily from your fuel emissions, is that the foremost vital human reason behind warming.

Greenhouse gas discharged every you burn one issue, be it associate automotive, plane or coal plant. This implies you wish to burn less fuel if you'd just like the Earth's climate to remain stable! And sadly, we have a bent to are presently destroying a variety of the best far-famed mechanisms for storing that carbon—plants. Deforestation can increase the severity of world warming additionally. Greenhouse gas is discharged from the human conversion of forests and grasslands into farmland and cities. All living plants store carbon. Once those plants die and decay, gas is discharged into the atmosphere. As forests and grasslands are live refined for your use, monumental amounts of carbon hold enter to the atmosphere. We can unstoppable this cycle. If the activities mentioned beyond heat the world just enough, it would cause natural carbon sinks to fail. A carbon sink may well be a natural system that stores carbon over thousands of years. Such sinks embrace human bogs and thus the arctic plain. But if these sinks destabilize, that carbon is discharged, in all probability inflicting unstoppable and harmful warming of the world. The oceans don't seem to be any more able to store carbon as they have in the past. The ocean may well be an oversized carbon sink, holding regarding 50 times the utmost quantity carbon as a result of the atmosphere. But presently scientists are live realizing that the inflated thermal stratification of the oceans has caused substantial reductions in levels of flora that store greenhouse gas. Inflated region carbon is to boot inflicting associate action of the ocean, since greenhouse gas forms acid once it reacts with water. The tiny plants of the ocean, the really bottom of that vast watery natural phenomenon, are laid low with the implications of world warming, which suggests they are turning into less able to store carbon, any tributary to action. As carbon sinks fail, the amount of carbon at intervals the atmosphere climbs (Venkataramanan and Smitha, 2011).

Anthropogenic greenhouse gas emissions have inflated since the pre-industrial era, driven for the foremost part by economic and increment and are presently more than ever. This has led to atmospherically concentrations of greenhouse gas, carbonic acid gas, and inhalation general an aesthetic that is unmatched exceedingly in a minimum of the last 800,000 years. Their effects, besides those of various phylogenies drivers, are detected throughout the climate system and are terribly doable to possess been the dominant reason behind the determined warming since the mid-20th century (Climate Amendment, 2014).

4.5   METHANE'S HUGE IMPACT

Methane series is created once being break down organic matter below oxygen-starved conditions. This happens once the organic matter is trapped underwater, as in rice paddies. It together takes place within the intestines of plant-eating animals, like cows, sheep, and goats. As a result of human, agriculture has big overtime to engulf most of the productive land in the world, it's presently adding many methane series to the atmosphere. Landfills and outpouring from fuel fields (methane may well be part of natural gas) are very important sources of methane series (Venkataramanan and Smitha, 2011).

Catharsis is live frozen chunks of ice and methane series that rest at an all-time low of the world's oceans. As a result of the water warms, the ice melts, and thus the methane series is discharged. If this warming that's caused by humans were to cause changes at intervals the Earth's ocean currents, then a quick melting of catharses would be potential. This too would turn out a regeneration loop that may cause any warming. It's believed that a variety of the warming cycles at intervals the Earth's history are caused by the sharp thawing of catharses (Venkataramanan and Smitha, 2011).

4.6   EFFECTS OF GLOBAL WARMING

Increasing international temperatures are live inflicting a broad of changes: Ocean levels are rising as a result of thermal growth of the ocean, boot to melting of land ice and changing of amounts and patterns of precipitation. The total annual power of hurricanes has already inflated markedly since 1975 as a result of their average intensity and average length have inflated (in addition, there has been a high correlation of cyclone power with tropical sea-surface temperature). Changes in temperature and precipitation patterns increase the frequency, duration, and intensity of various extreme weather events, like floods, droughts, heat waves, and tornadoes. Totally different effects of world warming embrace higher or lower agricultural yields, any glacial retreat, reduced summer stream flows, species extinctions (Venkataramanan and Smitha, 2011).

As an additional result of world warming, diseases like malaria infection are returning into areas where they need to be been extinguished earlier. So warming has effects on the amount and magnitude of these events, it's powerful to connect specific events to warming. So most studies specialize on the time up to 2100, while warming continues as a result of greenhouse gas has associate a calculable region lifetime of 50–200 years (Venkataramanan and Smitha, 2011).

4.7 EFFECTS ON WEATHER

Increasing temperature is maybe a lead guide to increasing precipitation but the implications on storms are live less clear. Any tropical storms partly depend on the gradient that's predicted to weaken at intervals the hemisphere as a result of the polar region warms quite the rest of the hemisphere. Regional effects of world warming vary in nature. Some are the results of a generalized international change, like rising temperatures, resulting in native effects, like melting ice. In numerous cases, associate change may even be related to associate change throughout a particular current or weather system. In such cases, the regional result may even be disproportionate and can't primarily follow the worldwide trend (Venkataramanan and Smitha, 2011).

There are live three major ways in that during which warming will produce changes to regional climate: melting or forming ice, dynamical the hydrological cycle (of evaporation) and dynamical currents at the oceans and airflow at the atmosphere. The coast can also be thought-about a zone, and might suffer severe impacts from water level rise (Venkataramanan and Smitha, 2011).

4.8 EFFECTS OF GLACIER RETREAT AND DISAPPEARANCE

Glacier retreat and disappearance: Mountain glaciers and snow cowl had attenuated in every the northern and southern hemispheres. This widespread decrease in glaciers and ice caps has contributed to determined water level rise (Venkataramanan and Smitha, 2011). Predictions are in connection with future changes in glaciers as:

- In Mountainous areas in Europe will face ice retreat.
- In Latin America changes in precipitation patterns and thus the disappearance of glaciers will significantly have an impact on water convenience for human consumption, agriculture, and energy production.
- In Polar Regions, there will be reductions in ice formation extent and thus the thickness of glaciers.

4.9   ROLE OF THE OCEANS

The role of the oceans in warming may well be an advanced one. The oceans perform a sink for gas, taking up galore that may otherwise keep at the atmosphere, but inflated levels of greenhouse gas have light-emitting diode to ocean action. What's a lot of, as a result of the temperature of the oceans can increase; they subsided able to absorb excess greenhouse gas. Warming is projected to possess type of effects on the oceans. Current effects embrace rising ocean levels as a result of thermal growth and melting of glaciers and ice sheets, and warming of the ocean surface, leading to inflated temperature stratification. Totally different potential effects embrace large-scale changes in ocean circulation (Venkataramanan and Smitha, 2011*).* Where Ocean warming dominates the increase in energy hold on at intervals the climate system, accounting for quite 90% of the energy accumulated between 1971 and 2010, with exclusively regarding 125% hold on at intervals the atmosphere. On a worldwide scale, the ocean warming is largest near the surface and thus the upper 75% heat by 1971 to 2010. It's nearly sure that the upper ocean (0.700 m) from 1971 to 2010.

Since the beginning of the industrial era, oceanic uptake of CO_2 has ended in acidification of the ocean; the pH of ocean surface water has reduced via 0.1, similar to a 26% boom in acidity, measured as hydrogen ion concentration. Over the length 1992 to 2011, the Greenland and Antarctic ice sheets have been dropping mass, probable at a larger rate over 2002 to 2011. Glaciers have persisted to shrink nearly worldwide. Northern hemisphere spring snow cover has continued to lower in extent. There is high confidence that permafrost temperatures have improved in maximum areas for the reason that early 80s in reaction to accelerated floor temperature and changing snow cover. The annual suggest Arctic sea-ice extent reduced over the duration 1979 to 2012, with a rate that become very probable inside the range 3.5–4.1% in step with decade. Arctic sea-ice quantities has reduced in every season and in every successive decade seeing that 1979, with the maximum rapid lower in decadal imply volume in summer. It is very probable that the annual suggest Antarctic sea-ice extent elevated within the range of 1.2 to 1.8% in keeping with decade among 1979 and 2012. However, there's excessive confidence that there are strong regional variations in Antarctica, with extent increasing in some regions and curtailmenting in others (Climate amendment, 2014).

4.10   ROLE OF CHLOROFLUOROCARBONS (CFCS)

CFCs have created every unnatural region cooling and warming supported these facts (Ashworth, 2016):

- CFCs destroyed gas at intervals the lower stratosphere/upper layer inflicting these zones at intervals the atmosphere to cool down 1.37°C from 1966 to 1998. This time span was chosen to eliminate the results of the natural star irradiance (cooling-warming) cycle result on the earth's temperature.
- The loss of gas allowed lots of ultraviolet light-weight UV no particulate radiation light to suffer the layer at a spare rate to heat the lower layer and 10" of the planet by 0.48°C (1966 to 1998).
- Mass and energy balances show that the energy that was absorbed at intervals the lower stratosphere/upper layer hit the lower troposphere/earth at a property level of $1.69 \times 10\ 18$ Btu more in 1998 than it did in 1966.
- Larger gas depletion at the Polar Regions caused these areas to heat up some a pair of and easy fraction (2 ½) times that of the everyday earth temperature (1.2°C versus 0.48°C). This has caused ground to melt that's cathartic copious quantities of methane series, calculable at 100 times that of manmade greenhouse gas unleash, to the atmosphere. Methane series at the atmosphere slowly converts to greenhouse gas and vapors and it's unleash has contributed to higher region greenhouse gas concentrations.
- There may well be a temperature anomaly in Antarctic continent. The Signe Island solid ground any north, warm just like the rest of the Polar Regions; but south at Votic, there has been a cooling result. Though' the cooling at Votic should be analyzed in further detail, as a results of the large hole there, black body radiation from Votic (some 11,400 feet above sea level) to location is presumptively the cause. Especially, since this development occurred over constant quantity that stratospheric gas destruction happened.

4.11   EFFECTS OF CO$_2$

From scientific analysis, the results of greenhouse gas ought to be really minimal as shown by degree earth temperature around 0.7 to 0.8°C from

1998 to 2008; greenhouse gas concentration at the atmosphere inflated some 20 ppm. It must be obvious to anyone that has analyzed action that greenhouse gas management the temperature, where CFC destruction of stratospheric gas correlates nicely with every the layer cooling and earth warming anomalies seen over the time span from 1966 to 1998. Although the atmosphere is tortuous in but it acts and reacts, CFCs appear to be the dominant reason behind larger than ancient earth warming. One can account for several, if not all of the 0.48°C rise in earth's temperature from 1966 to 1998 with the additional ultraviolet ray light-weight that hit the world as a result of the loss of gas at inter the layer. Unless the CFCs are aloof from the atmosphere and the earth, the whole earth would possibly still be hotter than ancient and higher concentrations of greenhouse gas (from ground unleash of methane) will exist until the CFCs at inter the layer slowly decrease naturally over future 50–100 years. The exceptions that may alter this are huge volcanic eruptions or weather modification techniques as projected by man of science citizen Dyson, whereby fine particulate, like mineral (Al_2O_3), is sprayed into the layer associated absorbs lots of ultraviolet ray light-weight to chill the planet just like the result seen with an oversize discharge. China and developing countries will terminate gas production. Some greenhouse gas production plants are finish up the previous schedule in developing countries and this could be really helpful. However, it'd not be that powerful to induce obviate existing CFCs from the layer and thus the earth's temperature will be brought back to ancient galore quicker (Ashworth, 2016). According to Einstein (Barnett, 1947), the grand aim of all science is to cover the simplest variety of empirical facts by logical deduction from the tiniest potential vary of hypotheses or axioms. One can do this here to destruction of greenhouse gas to elucidate all of the determined recent earth and atmosphere temperature anomalies from 1966 to 1998.

4.12   IMPACTS OF CLIMATE CHANGE

In recent decades, changes in climate have caused impacts on natural and human systems on all continents and across the oceans, that Impacts are as results of determined action, no matter its cause, indicating the sensitivity of natural and human systems to dynamical climate. Proof of determined climate change impacts is the strongest and most comprehensive for natural systems. In many regions, dynamical precipitation or melting snow and ice are live fixing hydrological systems, poignant water resources in terms of quantity and quality. Many terrestrial, contemporary, and marine species have shifted their geographic ranges, seasonal activities, migration patterns, abundances, and species interactions in response to the current change in climate change. Some impacts on human systems have together been attributed to climate change, with a major or minor contribution of temperature change distinguishable from totally different influences. Assessment of the various studies covering an oversized vary of regions and crops shows that negative impacts of climate change on crop yields are lots of common than positive impacts. Some impacts of ocean action on marine organisms are attributed to human influence (Climate Amendment, 2014).

4.13 FUTURE CLIMATE CHANGES: RISKS AND IMPACTS

Cumulative emissions of the greenhouse, for the most part, confirm international mean surface warming by the late twenty-first century. And on the far side, projections of greenhouse gas emissions vary over an oversized vary, counting on every socio-economic development and climate policy, evolution GHG emissions are live primarily driven by population size, economic activity, lifestyle, energy use, land use patterns, technology, and climate policy. The Representative Concentration Pathways (RCPs), that are a measure used for making projections supported these factors, describe four utterly different twenty-first-century pathways of GHG emissions and region concentrations, air waste emissions and land use. The RCPs embrace a stern mitigation state of affairs, two intermediate things and one state of affairs with terribly high GHG emissions (Climate Amendment, 2014).

Future climate will depend on committed warming caused by past phylogenies emissions, additionally as future phylogenies emissions and natural climate variability. The worldwide mean surface natural process for the amount 2016–2035 relative to 1986–2005 is comparable for the four RCPs and might doable be at intervals the vary 0.3°C to 0.7°C. This assumes that there will be no major volcanic eruptions or changes in some natural sources (e.g., CH_4 and N_2O), or sharp changes in total star irradiance. By the mid-21st century, the magnitude of the projected change is significantly full of the choice of emissions state of affairs. Relative to 1850–1900, the international surface natural process for the tip of the twenty-first century (2081–2100) is projected to doable exceed 1.5°C for RCP 4.5, RCP 6.0 and RCP 8.5. Warming is maybe visiting exceed 2°C for RCP 6.0 and RCP 8.5, lots of doable than to not exceed 2°C for RCP 4.5, but unlikely to exceed 2°C for RCPs. The rise of world mean surface temperature by the tip of the twenty-first century (2081–2100) relative to 1986–2005 is maybe going to be 0.3°C to 1.7°C below RCP 2.6, 1.1°C to 2.6°C below RCP 4.5, 1.4°C to 3.1°C below RCP 6.0 and 2.6°C to 4.8°C below RCP 8.59. The Arctic region will still heat quicker than the worldwide mean. It's nearly sure that there will be lots of frequent hot and fewer cold temperature extremes over most land areas on daily and seasonal timescales, as the international mean surface temperature can increase. It's really doable that heat waves will occur with the subsequent frequency and longer length. Occasional cold winter extremes can still occur (Climate Amendment, 2014).

4.14   FUTURE RISKS AND IMPACTS CAUSED BY A CHANGING CLIMATE

Risk of climate-related impacts results from the interaction of climate-related hazards (including risky events and trends) with the vulnerability and exposure of human and natural systems, likewise as their ability to adapt. Rising rates and magnitudes of warming and totally different changes at intervals the climate system, among ocean action, increases the danger of severe, pervasive and in some cases irreversible prejudices impacts. Some risks are live considerably relevant for individual regions, whereas others are international. The overall risks of future impacts are typically reduced by limiting the speed and magnitude of global climate change, likewise as ocean action. The precise levels of worldwide (climate amendment global climate change temperature change) spare to trigger abrupt and irreversible change keep unsure, but the danger associated with crossing such thresholds can increase with rising temperature. For risk assessment, it is necessary to evaluate the widest potential vary of impacts, likewise as low-probability outcomes with huge consequences.

Currently, weather change is the greatest danger to life on Earth. Recent studies show the 20 warmest years on record have been in the beyond 22 years, and the top four inside the beyond four years. Over the past century, the average temperature of the Earth has risen by 1.8°F. Over the subsequent 100 years, scientists are projecting another 0.5 to 8.6°F rise in the temperature. The cause of this temperature change is human activities which have released massive amounts of carbon dioxide and different greenhouse gases (such as methane, nitrous oxide and fluorinated gases) into the atmosphere. The majority of greenhouse gases come from burning fossil fuels to produce energy. Results of this worldwide warming include rising ocean temperatures, a greater acidic ocean, melting ice caps, and growing sea degrees.

Recent studies display that Earth has lost 1/2 its wildlife in just 40 years. The IUCN expected that 20–30% of plant and animal species are likely to be at an increased chance of extinction because of weather change. Climate change will also affect human beings and mainly the world's 2.6 billion poorest people. According to the United Nations Development Program, "Receding forests, changing rainfall patterns and rising sea levels will exacerbate current economic, political and humanitarian stresses and have an effect on human development in all parts of the world." Everyone can take easy actions to lessen greenhouse gas emissions, mitigate our personal contribution to climate change, and build more and greater community resiliency.

4.15 CLIMATE CHANGE BEYOND 2100: IRREVERSIBILITY AND ABRUPT CHANGES

Many aspects of climate change and associated impacts will continue for many years, nonetheless, phylogenies emissions of greenhouse gases are live stopped. The risks of abrupt or irreversible changes increase as a result of the magnitude of warming can increase. Warming will continue on so many facets of 2100. Surface temperatures will keep nearly constant at elevated levels for many centuries once an entire halt of greenhouse gas emissions. An oversized fraction of phylogenies climate change succeeding from greenhouse gas emissions is irreversible on a multi-century to millennial timescale, except in the case of an oversized internet removal of greenhouse gas from the atmosphere over a sustained quantity. The stabilization of the world average surface temperature does not imply stabilization for all aspects of the climate system. Shifting biomes, soil carbon, ice sheets, ocean temperatures, and associated water level rise all have their own intrinsic long timescales which might finish in changes lasting many to thousands of years once the international surface temperature is stable. There's high confidence that ocean action will increase for many years if greenhouse gas emissions continue, and might powerfully have an impact on marine ecosystems.

It is nearly sure that international mean water level rise will continue for many centuries on the so many facets up to 2100, with the quantum of rising smitten by future emissions. The sting for the loss of the Greenland ice sheet over a millennium or lots of, associate degreed associate associated water level rise of up to seven m, is larger than regarding 1°C but however concerning 4°C of world warming with reference to pre-industrial temperatures. Abrupt and irreversible ice loss from the Antarctic ice sheet is possible, but current proof and understanding keep to make a quantitative assessment. Magnitudes associate degreed rates of worldwide climate amendment associated with medium- to high-emission things cause associate inflated risk of abrupt and irreversible regional-scale change at intervals the composition, structure, and performance of marine, terrestrial, and contemporary ecosystems, likewise as wetlands. A reduction in ground extent is simply regarding sure with a continuing rise in international temperatures (Climate Amendment, 2014).

4.16   HOW CAN WE DETERMINE FUTURE CLIMATE?

The challenges of predicting weather and climate are live really utterly totally different. Predicting the weather is like predicting but a particular eddy will move and evolve throughout a turbulent river: it's potential over short time scales by extrapolating the previous path of the eddy, but eventually, the eddy is influenced by neighboring eddies and currents to the extent that predicting its actual path and behavior becomes unfeasible. Similarly, the limit for predicting individual weather systems at intervals the atmosphere is around 10 days. On the alternative hand, predicting climate is like predicting the flow of the whole watercourse. It necessarily a concept of the most forces dominant the watercourse like changes in downfall, the operation of dams, and extraction of water. Projections of human-induced action over decades to centuries are live potential as a result of human activities have sure effects on the long-term region composition, and in turn a sure result on climate (Holmes, 2015). Figure 4.2 shows the irrigation canal in Upper Egypt.

FIGURE 4.2   Irrigation canal, Upper Egypt (Hamada, 2018).

4.17   HOW DO WE DETECT CLIMATE CHANGE?

Identifying the natural process that is international in extent desires frequent observations from many locations around the world. Thermometers, rain gauges, and totally different simple instruments are accustomed to live climate variables, starting at intervals in the mid-19th century. Over time the quality, choice, and quantity of observations have improved. Since the 70s, refined sensors on earth-orbiting satellites have provided near international coverage of the various climate variables. By strictly analyzing the data gathered by techniques (with careful account for changes in instrument varieties, experimental practices, instrument locations, and concrete areas), it has been potential to map the distribution of temperature and totally different climate changes since the late 19th century. To study climate changes that occurred before direct measurements were created, scientists use proof from totally different sources that record a climate signal. These embrace climate signals encoded at the composition of ice cores, corals, sediments in oceans and lakes, and tree rings. Of those records are measure ordered down consecutive overtime as sediments accumulate. Ice cores from polar ice sheets, that are live built from snow ordered down over tens to several thousands of years, supply records of every past greenhouse gas and temperature. As a result of the snow transforms into ice, it traps air in sealed bubbles that provide a sample of past region composition, whereas the relation of stable isotopes of either gas or part at intervals the water molecule is claimed to the temperature at the time once the snow fell. A lot of trendy historical changes are typically glorious by analyzing written and pictorial records, as an associate example of changes in formation extent (Holmes, 2015).

4.18   GLOBAL WARMING AS A BOON AND A BANE

With all the confusion encompassing the worldwide warming discussion, galore of the overall public exclusively takes from the media that warming will hurt our world. However, warming has potential advantages additionally as drawbacks. Initial of all, as a result of the temperature, can increase, the world will have an associate extended season in many areas. In general, there will be less chilling weather, and thus the inflated temperatures and gas levels will alter lots of plant growth. With lots of plants, growth associated an extended season; there will be lots of food for people. The warmer weather additionally can utterly have an impact on transportation. Airplanes, trains, buses, and cars will stop having cold weather-related delays for ice and snow. Thus, contrary to common belief, warming can have some advantages (Moore, 2008).

Of course, warming has many negative effects on the world. Warming is and might still have dramatic effects on aquatic life and variety. To compound the effects of the natural prejudice to ecosystems associated humans would possibly any disturb the ecosystems? As an example, by creating an effort to combat the implications of Rising Ocean levels, man could suitable continue the lineation. In our tries to defend the lineation surroundings, totally different species would possibly disappear. Inflated temperatures additionally can negatively have an impact on the food supply in many places, nullifying the benefits of an associate extended season. The temperature can increase will bring hotter temperatures at intervals the summer, which might cause plants to die. It would together cause weather patterns, sort of a lot of intense floods and storms (Moore, 2008). Within the future, warming would possibly encourage be a boon and a nemesis to the world, but it isn't doable to estimate merely what amount of either it will be.

4.19 CONCLUSION

According to NASA and IPCC, Global temperature has increased through 1.4 of since 1880, CO_2 stages has reached 400.71 parts per billion, lack of world's forest cover among the duration 2000 and 2012 is 1.5 million square km, reduction of land ice 287 billion metric ton per year, sea level rise is 3.2 mm per year and lack of arctic ice cover at the rate of 13.3% per decade. Increasing threat of irreversible changes due to large scale shift within the climate system together with several touchy species which includes ocean corals, aquatic birds, reptiles such as sea turtles and amphibians are facing extinction, failing of crops cause famine in lots of East African countries, decrease in potable water in Mediterranean and Southern Africa and growing intensity of severe activities which includes wooded area fires (Australia and Indonesia), flooding (Bangladesh) , storm activities (tornadoes and hurricanes in USA), droughts (Sahel region) and deadly heat waves (in India 2015) recorded in lots of parts of the world. Anthropogenic release of greenhouse gases CO_2, CH_4, water vapors, N_2O, O_3, HFCs, PFCs and SF6⁻reflects a portion of solar power lower back to the earth, this increases the temperature, causes adjustments in ocean currents, seasonal weather styles and in the long run adjustments the weather. Deforestation reduces the CO_2 sink and it similarly comple-ments the greenhouse effect. Several mitigation methods including use of alternative green energy sources, decreasing the use of fossil fuels, use of greenhouse gasoline reduction techniques for the duration of the emission, carbon capture & carbon sequestration, afforestation, reforestation, protection of current forest reserves, silviculture and agroforestry are being facilitated by numerous international, authorities and non-governmental organizations. Climate change issue may be treated both adapting to the change or disaster danger reduction. UNDP has suggested a three step approach to work on Carbon finance consist of removal of limitations to climate friendly technologies, establishing efficient host country techniques for clean development mechanism (CDM) and expand projects through millennium development goal (MDG) carbon facility. An Integrated Territorial Climate Plan (ITCP) becomes designed for local governments to plan their activities along with financing weather change mitigation process.

KEYWORDS

- **carbonic acid**
- **environmental stresses**
- **global warming**
- **greenhouse gases (GHGs)**
- **Kyoto protocol (KP)**
- **tackle heating**

REFERENCES

Andrew, H., (2015). *The Science of Climate Change Questions and Answers*. Australian Academy of Science. ISBN: 978-0-85847-413-0. www.science.org.au/climatechange (accessed on 4 January 2020).

Barnett, L., (1947). *The Universe and Dr. Einstein* (p. 110). William Sloane Associates, New York.

Climate Change, (2014). *Synthesis Report: Summary for Policymakers*. The IPCC's Fifth Assessment Report (AR5).

Elasha, B., et al., (2006). *Background Paper on Impacts: Vulnerability and Adaptation to Climate Change in Africa*. Paper presented for the African workshop on adaptation implementation of decision 1/CP.10 of the UNFCCC convention, Accra, Ghana. Online:

http://unfccc.int/files/adaptation/adverse_effects_and_response_measures_art_48/application/pdf/200609_background_african_wkshp.pdf (accessed on 4 January 2020).

FAO, (2006). Food and Agriculture Organization. *The State of Food Insecurity in the World*. Online: http://www.fao.org/docrep/009/a0750e/a0750e00.htm (accessed on 4 January 2020).

Hamada, Y. M. (2018). Special pictures album, Took by Youssef M. Hamada, Egypt, 2018.

IPCC, (2007). Intergovernmental panel on climate change. *Climate Change 2007: Climate Change Impacts*. Adaptation and vulnerability, IPCC WGII fourth assessment report. Online: http://www.ipcc.ch/SPM6avr07.pdf (accessed on 4 January 2020).

Moore, T., (2008). *EMBO Reports, 9*, S41–S45.

Robert A. Ashworth (2016). Ozone destruction major cause of warming! – Part 2. *Hydrocarbon Processing 88*(11), November 2009 with 18 Reads.

Venkataramanan, M., & Smitha, (2011). Causes and effects of global warming. *Indian Journal of Science and Technology, 4*(3), ISSN: 0974–6846.

CHAPTER 5

Climatic Impacts on Crop Yields and Net Farm Incomes

ABSTRACT

Climate change and agriculture are units indissolubly connected. Agriculture still depends altogether on the weather. Temperature change has previously created a negative impact on agriculture in several zones of the world due to additional stark weather patterns. Temperature change is determinable to still as a result of worsen geologic process, floods, and disrupt growing seasons. The Food and Agriculture Organization (FAO) warns that an increase in average international temperatures of 2–4°C over pre-industrial levels would possibly reduce crop yields by 15–35% in Africa and western Asia, and by 25–35 within the Middle East. An increase of 2°C alone would possibly presumptively cause the extermination of millions species (Ellis, 2008). Figure 5.1 shows clouds in summer as evidence of climate change in Egypt.

Agricultural practices conjointly irritate temperature change. The Intergovernmental Panel on Climate Change (IPCC) says that agriculture contributes 13.5% of global greenhouse gas emissions (2004). As claimed by Greenpeace, if calculate each the direct and indirect emissions from the food system, agriculture's contribution is also as high as 32% (according to Greenpeace, includes all associated activities; in addition to agricultural production, they add land use, transportation, packaging, and processing). The future of agricultural production depends on each designing new ways in which to adapt to the in all probability consequences of temperature change, additionally as slashing agricultural practices to mitigate the climate injury that current practices cause, all without undermining food security, rural development, and livelihoods. Usually, this can be often an oversized endeavor. Global climate change and food security (CCFS) are connected as climate change can directly have a sway on a country's ability to feed its people. This chapter shows the matter of climatic impacts on crop yields and net farm incomes through the views of the foremost specialists in climatic impacts on crop yields and farm income.

FIGURE 5.1   Clouds in summer as evidence of climate change, Egypt (Hamada, 2018).

5.1   BACKGROUND

Agriculture is an economic activity that is extraordinarily dependent upon weather and climate therefore to provide the food and fiber necessary to sustain human life. Not surprisingly, agriculture is deemed to be associate degree economic activity that is expected to be at risk of climate variability and alter. The vulnerability of agriculture to climate variability and alter could be a problem of major importance to the international scientific community, and this concern is reflected in article one of the UNFCCC, that demand the stabilization of greenhouse gas concentrations at intervals the atmosphere at level that, would forestall serious evolution interference with the climate system. Such level must be achieved within a timeframe adequate to: (i) change ecosystems to adapt naturally to climate change; (ii) confirm that food production is not threatened; and (iii) change economic development to proceed in an exceedingly property manner (IISD and EARG, 1997). On a world basis, climate variability and alter may have an overall negligible impact on total food production; however, the regional impacts are in all probability to be substantial and variable, with some regions taking advantage of associate degree altered climate and completely different regions adversely affected. Generally, food production is maybe visiting decline in most significant regions (e.g., subtropics and tropical areas), wherever as agriculture in developed countries might fine profit where technology is further offered and applicable accommodative changes are utilized (IISD and EARG, 1997).

In terms of endeavor a national assessment of global climate change impacts on African agriculture, this sector has received extended attention relative to completely different resource sectors. There have been a significant variety of analyses and studies directed at temperature change and agriculture in the continent, and much of this literature has already been compiled in annotated list kind (Wheaton, 1994). Further, agriculture has received attention on a regional basis (e.g., Projects the MBIS and the GLSLB), thus the projected tract Climate Impact and Adaptation Study, and some areas, like parts of the Prairies, are examined in terms of its palace record (the Palliser Triangle Study), and current and future climate (the Nat Dame Agatha Mary Clarissa Christie Foundation Project in Alberta).

Thought here's a substantive body of literature focusing upon temperature change and agriculture, there continues to be a decent vary of opinions referring to things of future impacts. At intervals the Prairies, for example, that's associate degree agricultural area of significant international, national, and regional importance; the foremost recent IPCC Assessment Report given a cautiously optimistic and but foreboding appraisal: accumulated agricultural production on the Prairies is also an opportunity with higher temperatures and dioxide levels, provided adaptation measures unit of measurement undertaken and adequate downfall happens. But this appraisal, it's easy that grassland agriculture has had an actual history in terms of its vulnerability to the vagaries of climate. The drought of the 30s, and last in 1988, illustrates the sensitivity of agriculture throughout this region to severe wet deficits. Combined with recent floods, the ever-presence of hail injury, and completely different climatic events, the story and current conditions underscore the variability of climate and thus the negative effects from extreme weather events (IISD and EARG, 1997).

Given the historical sensitivity of grassland agriculture to climate, the uncertainty that exists in our things, and thus the economic and social importance of this sector, there is a strong ought to improve our understanding of temperature change impacts and thus the flexibility of grassland agriculture. As a primary step to handle this wish, the aim of this section is to gift a quick summary of the literature on agriculture and temperature change impacts and adaptation with it in the Prairies (IISD and EARG, 1997).

5.2 SCOPE OF THE CHAPTER

There is extended support for organic farming because it is the best method to mitigate greenhouse gas. Organic agriculture's emissions are typically below those of business agriculture. Though some modes of organic agriculture do not harvest yields as high as industrial or chemical agriculture, it is a further property suggests that of cultivating the land. It builds soil quality and uses further various cropping systems that in turn cut back the amount of greenhouse gases emitted. And it's higher at sequestering (absorbing) carbon and chemical element than industrial agriculture (Ellis, 2008).

There is, however, mixed views referring to that model of agriculture most effectively minimizes temperature change. In some situations, industrial agriculture produces fewer greenhouse emissions. There is nearly universal agreement among researchers, however, that the impact of chemical fertilizers is negative. Opinions vary greatly referring to the amounts of energy utilized in fully completely different production processes and concerning that, new techniques most effectively minimize greenhouse emissions. Organic standards haven't typically developed climate-related criteria (e.g., dominant what amount oil is utilized on-farm in production and harvest of crops). Further comparative analysis should be conducted on industrial versus organic and small-scale versus large-scale farming. Of the points reviewed, the Greenpeace paper makes the foremost daring arrange to answer these needs. The analysis should deepen, there is jointly a growing body of analysis shows, but organic farming can feed the world and that change from chemical agriculture to organic agriculture would not reduce the world's food provide (Ellis, 2008).

5.3 MANY COMPLEX PROCESSES SHAPE OUR CLIMATE

Based precisely on the physics of the amount of energy that carbon dioxide absorbs and it emits, a doubling of atmospherically carbon dioxide on centration from pre-industrial levels (up to concerning 560 ppm) would, by itself, cause a world average temperature increase 1°C (1.8°F). At the general climate system, things are further complex; warming results in further effects (feedbacks) that either amplify or diminish the initial warming. The foremost very important feedbacks involve various forms of water. A warmer atmosphere typically contains further vapors. Vapor is also a potent greenhouse gas emission, therefore inflicting further warming; its short life in the atmosphere keeps its increase for the foremost half in step with warming. Thus, the vapor is treated as equipment, and not a driver of climate change. Higher temperatures within the Polar Regions soften ocean ice and cut back seasonal snow cowl, exposing a darker ocean and land surface which can absorb further heat, inflicting further warming. Another very important but unsure feedback, problems change in clouds. Warming can increase vapor and may cause overcasts to increase or decrease which can either amplify or dampen activity betting on the changes at intervals the horizontal extent, altitude, and properties of clouds. Figure 5.2 shows warming and increases in water vapor together may cause cloud in Upper Egypt, Egypt. The most recent assessment of the science indicates that the international impact of clouds amendment is maybe going to amplify warming (Cicerone and Nurse, 2016).

FIGURE 5.2 Warming and increases in water vapor together may cause clouds, Egypt (Hamada, 2018).

The ocean moderates climate change. The ocean is also an oversized heat reservoir, but it's powerful to heat its full depth as a result of heat water tends to stay near the surface. The speed at that heat is transferred to the deep ocean is slow; it varies from year to year and from decade to decade, and helps to determine the pace of warming at the surface. Observations of the sub-surface ocean are restricted before regarding to 1970; but since then, warming of the upper 700 m (2,300 feet) is quickly apparent. There is jointly proof of deeper warming. Surface temperatures and downfall in most regions vary greatly from the worldwide average due to geographical location, specifically latitude and continental position, every the standard values of temperature, rainfall, and their extremes (which typically have the largest impacts on natural systems and human infrastructure), are also powerfully stricken by native patterns of winds (Cicerone and Nurse, 2016).

Estimating the implications of feedback processes, the pace of the warming, and regional temperature change needs the employment of mathematical models of the atmosphere, ocean, land, and ice (the cry sphere) designed upon established laws of physics and thus the most recent understanding of the physical, chemical, and biological processes moving climate, and run on powerful computers. Models vary in their projections of what amount additional warming to expect (depending on the sort of model and on assumptions utilized in simulating sure climate processes, notably cloud formation and ocean mixing), but all such models agree that the impact of feedbacks is to amplify warming (Cicerone and Nurse, 2016).

5.4   HUMAN ACTIVITIES ARE CHANGING THE CLIMATE

Rigorous analysis of all data and rules of proof shows that all of the worldwide determined warming up over the past 50 years about can't be explained by natural causes and instead requires a significant role in the influence of human activities. Therefore, on distinguishing the human influence on climate, scientists ought to take under consideration many natural variations that have a sway on temperature, precipitation, and completely different aspects of climate from native to an international scale, on time scales from days to decades long-terms. One natural variation is that, the El Niño southern oscillation (ENSO), an irregular alternation between warming and cooling (lasting two to seven years) in the equatorial Pacific Ocean that causes vital from year-to-year regional and global shifts in temperature and rainfall patterns. Volcanic eruptions alter climate, partly increasing the amount of very little (aerosol) particles at the layer that mirrors or absorb daylight, leading to a short surface cooling lasting typically concerning two to three years. Over several thousands of years, slow, revenant variations in Earth's orbit around the Sun, that alter the distribution of alternative energy received by Earth, are enough to trigger the geological period cycles of the past 800,000 years (Cicerone and Nurse, 2016).

Fingerprinting is also a robust methodology of learning the causes of climate change. Completely different influences on climate cause different patterns in climate records. This becomes obvious once scientists probe on the so much facet changes at intervals the common temperature of the planet and look further closely at geographical and temporal patterns of climate change. For example, an increase at intervals the Sun's energy output may end up in an exceedingly very fully completely different pattern of climate change (across Earth's surface and vertically at intervals the atmosphere) compared to it evoked by an increase in carbon dioxide concentration. Determined region temperature changes show a fingerprint copious nearer thereto of a long-term carbon dioxide increase than to it of an unsteady Sun alone. Scientists routinely take a glance at whether or not strictly natural changes at intervals the Sun, volcanic activity, or internal climate variability would possibly in all probability build a case for the patterns of change they have determined in many alternative aspects of the climate system. These analyses have shown that the determined climate changes of the past several decades can't be explained just by natural factors (Cicerone and Nurse, 2016).

5.5 IF THE WORLD IS WARMING, WHY ARE SOME WINTERS AND SUMMERS STILL VERY COLD?

Global warming is also a long-term trend, but that does not mean that every year is going to be hotter than the previous one. Day to day and year to year changes in weather patterns will still manufacture some remarkably cold days and nights, and winters and summers, even as the climate warms. Figure 5.3 shows why are some winters and summers still very cold in Egypt?

Climate change suggests that not only changes in globally averaged surface temperature, but jointly changes in region circulation, at intervals the scale and patterns of natural climate variations, and in native weather. El Niño events shift weather patterns so that some regions are created wetter, and wet summers are typically cooler. Stronger winds from Polar Regions can contribute to an occasional colder winter. In an exceedingly similar methodology, the persistence of part of a region circulation pattern referred to as the Atlantic Ocean Oscillation has contributed to several recent cold winters in Europe, Japanese North America, and Northern Asia (Cicerone and Nurse, 2016).

Atmospheric and ocean circulation patterns will evolve as Earth warms and may influence storm tracks and many of various aspects of the weather.

FIGURE 5.3   Why some are winters and summers still very cold in Egypt (Hamada, 2018).

Heating tilts the chances in favor of further warm days and seasons and fewer cold days and seasons. For example, across the US in the 1960s, there are further daily record low temperatures than record highs, but in the 2000s, there are over doubly as many record highs as record lows. Another main example of tilting the chances is that over recent decades heat waves have accumulated in frequency in large parts of Europe, Asia, and Australia (Cicerone and Nurse, 2016).

5.6   WHY IS ARCTIC SEA ICE DECREASING WHILE ANTARCTIC SEA ICE IS NOT?

Sea ice extent is affecting from winds and ocean currents additionally as temperature. Ocean ice at intervals the partly-enclosed ocean seems to be responding on to warming, whereas amendments in winds and at intervals the ocean seem to be dominating the patterns of climate and ocean ice amendment at intervals the ocean around Antarctic continent. Ocean ice at the Arctic has reduced dramatically since the late 70s, notably in summer and autumn. Since the satellite record began in 1978 (being for primary time whole and continuous region coverage of the Arctic), the yearly minimum Arctic Ocean ice extent (which happens in early to mid-September) has reduced by over 40%. Ice layer expands once more each Arctic winter but the ice is diluting than it will not be. Estimates of Past Ocean ice extents counsel that this decline may even be new in an exceedingly minimum of the past 1,450 years. The complete volume of ice, the merchandise of ice thickness and area, has reduced faster than the ice extent over the past decades. Since ocean ice is extremely reflective, warming is amplified as a result of the ice decreases and extra sunshine is absorbed by the darker underlying ocean surface (Cicerone and Nurse, 2016). Figure 5.4 shows the ice layer expands again this winter in New Cairo in Egypt however, the ice is diluting in the Antarctic.

FIGURE 5.4   Ice layer expands this winter in New Cairo, Egypt (Hamada, 2018).

Sea ice at the Antarctic has shown a small increase in extent since 1979 overall, though some areas, like that to the west of the Antarctic Peninsula, have versed a decrease. Changes in surface wind patterns around the continent have contributed to the Antarctic pattern of ocean ice change whereas ocean factors just like the addition of cool water from melting ice shelves may also have competing for a task. The wind changes embody a recent strengthening of westerly winds that reduces the amount of warm air from low latitudes penetrating into the southern high latitudes and alters the tactic at intervals that ice moves aloof from the continent. The change in winds may finish partially from the implications of stratospheric gas depletion over the Antarctic continent (i.e., the hole, a development that is distinct from the human-driven changes in lasting greenhouse gases). However, short trends at intervals the Southern Ocean, like those determined, can promptly occur from the natural variability of the atmosphere, ocean, and ocean ice system (Cicerone and Nurse, 2016).

5.7 HOW DOES CLIMATE CHANGE AFFECT THE STRENGTH AND FREQUENCY OF FLOODS, DROUGHTS, HURRICANES, AND TORNADOES?

Earth's lower atmosphere is turning into hotter and moister as a result of human-emitted greenhouse gases. This provides the potential for extra energy for storms and sure severe weather events, per theoretical expectations, serious downfall and snow events (which increase the danger of flooding) and heat waves are typically turning into further frequent. Trends in extreme downfall vary from region to region: The foremost-pronounced changes are evident in North America and parts of Europe, significantly in winter. Trends in extreme weather events to climate change are troublesome as a result of these events are by definition rare and therefore exhausting to gauge faithfully, and are stricken by patterns of natural climate variability. As an example, the foremost vital reason for droughts and floods around the world is that the shifting of climate patterns between El Niño and La Niå events. On land, El Niño events make drought in many tropical and subtropics areas, whereas La Niå events promote wetter conditions in many places, as went on in recent years. These short and regional variations are expected to become additional extreme throughout a warming climate (Cicerone and Nurse, 2016). Figure 5.5 shows how does climate change affects the strength and frequency of floods, droughts, hurricanes, and tornadoes?

FIGURE 5.5   How does climate change affect the strength and frequency of floods, droughts, hurricanes, and tornadoes (Hamada, 2018)?

There is substantial uncertainty concerning hurricanes. The impact of climate change on cyclone frequency remains a difficulty of current studies. Whereas changes in cyclone frequency keep unsure, basic understanding and model results advocate that, the strongest hurricanes (when they occur) are in all probability to become further intense and presumptively larger throughout a warmer, moister atmosphere over the oceans. Usually this can be} often supported by offered empirical proof at intervals the Atlantic Ocean. Some conditions favorable for strong hinder storms that spawn tornadoes are expected to increase with warming; but uncertainty exists in numerous factors that have a sway on tornadoes formation, like changes at intervals the vertical and horizontal variations of winds (Cicerone and Nurse, 2016).

5.8   HOW FAST IS SEA LEVEL RISING?

Long-term measurements of tide gauges and the new satellite data show that international water level is rising, with best estimates of the global-average rise over the last 20 years centered on 3.2 millimeters annually

(0.12 inches per year). The determined rise since 1901 is concerning 20 centimeters (8 inches). Observations show that the worldwide average water level has up by concerning 20 centimeters (8 inches) since the late 19th century. The water level is rising faster in recent decades; measurements from tide gauges (blue) and satellites (red) indicate that the foremost effective estimate for the standard water level rise over the last 20 years is concentrated on 3.2 millimeters annually (0.12 inches per year) (Cicerone and Nurse, 2016).

This low-lying rise has been driven by (in order of importance): the growth of water volume as a result of the ocean warms, melting of mountain glaciers in most regions of the world, and losses from the Greenland and Antarctic ice sheets. All of those result from a warming climate. Fluctuations in water level jointly occur as a result of changes at the amounts of water hold on toward the shore. The amount of ocean level change at any given location jointly depends on a variety of various factors, likewise as whether or not regional earth science processes and rebound of the land laden by previous ice sheets are inflicting the land itself to rise, and whether or not changes in winds and currents are pillar ocean water against some coasts or moving water away (Cicerone and Nurse, 2016).

The effects of rising water levels are felt most acutely at intervals the accumulated frequency and intensity of occasional storm surges. If carbon dioxide and completely different greenhouse gases still increase on their current trajectories, it's projected, that water level may rise by an extra 0.5 to at least one m (1.5 to 3 feet) by 2100. But rising ocean levels will not stop in 2100; ocean levels are going to be copious higher at intervals the subsequent centuries as a result of the ocean continues to require up heat and glaciers still retreat. It remains powerful to predict the most points of but the Greenland and Antarctic Ice Sheets will reply to continued warming, but it's thought that Greenland and will be West Antarctic continent will still lose mass, whereas the colder Greenland of Antarctic continent would possibly begin to comprehend, as they receive snowfall from hotter air that contains further wet. Water level at the last interglacial (warm) quantity around 125000 years gone peaked at all told chance 5 to 10 meters over this level. Throughout this era, the Polar Regions were hotter than they are nowadays. This suggests that, over millennia, long periods of accumulated heat may end up in the very important loss of parts of the island and Antarctic Ice Sheets and to resultant water level rise (Cicerone and Nurse, 2016). Figure 5.6 shows the average rains in Africa.

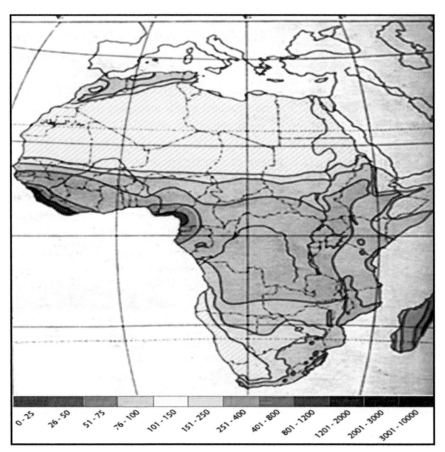

FIGURE 5.6   Average rains, Africa (Hamada, 2018).

5.9   WHAT IS OCEAN ACIDIFICATION AND WHY DOES IT MATTER?

Direct observations of ocean chemistry have shown that the beam balance of ocean water has shifted to an extra acidic state (lower pH). Some marine organisms (such as corals and some shellfish) have shells composed of carbonate that dissolves further promptly in acid. As a result of the acidity of ocean water that can increase, it becomes more durable for them to create or maintain their shells (Cicerone and Nurse, 2016). Figure 5.7 shows why are some summers still very cold in North Egypt?

FIGURE 5.7 Why are some summers still very cold in North Egypt (Hamada, 2018).

Carbon dioxide gas dissolves in water to create a weak acid, and thus the oceans have absorbed a couple of third of the carbon dioxide ensuing from human activities, resulting in a gradual decrease in ocean pH levels. With increasing atmospheric carbon dioxide gas, analytical balances can amendment even additional throughout the subsequent century. Laboratory and completely different experiments show that at a lower place high dioxide and in further acidic waters, some marine species have ill-shapen shells and lower growth rates, though the impact varies among species. Natural action conjointly alters the sport of nutrients and lots of different components and compounds within the ocean, and it's doubtless to shift the competitive advantage among species, with as-yet-to-be-determined impacts on marine ecosystems and also the organic phenomenon (Cicerone and Nurse, 2016).

5.10   FUTURE IMPACTS OF CLIMATE CHANGE ON DEVELOPING COUNTRIES

Depending on emission screenplays and models used, the temperature rise is determinable to be between 1–6°C at intervals the subsequent century (likely to be between 2–4°C). The value of warming throughout the last consecutive century mainly depends on the speed of increase in greenhouse gas emissions. To remain temperature rise within 2°C—a vital of EU targets—significant action to chop back emissions is needed. The international temperature increase will not be uniform. Regions around the poles will see further warming than tropical regions (Ludwig et al., 2007).

The most impacts of the world climate amendment will not be felt through higher temperatures but through a change at intervals the hydrological cycle. Heating will intensify the hydrological cycle that is ready to extend international precipitation. Changes in rainfall downfall patterns will not be equally unfold around the globe. Around the tropics and at the northern part of the hemisphere (Canada, Russia, and Northern Europe) rainfall is maybe going extend, however in most of the sub-tropical regions—where many developing countries are placed – rainfall is probably going to scale back (Ludwig et al., 2007).

Rainfall is projected to change around the globe. Briefly, the following changes are predicted for developing countries: Southern Africa is probably going to drier; East Africa and also the Horn of Africa can probably receive additional rainfall downfall. For the Sahel, the changes are still unclear. In Latin America, the Caribbean, the Amazon, and Chile are in all probability to check a discount of rainfall downfall. For the South-Eastern part of South America rise in summer, the downfall is predicted. In Asia, the Indian earth would possibly see an increase of precipitation throughout the monsoon season but the lower downfall is predicted outside the season. This suggests that the variations between the seasons will increase, with additional drought stress throughout the time of the dry season, as results of higher temperatures and fewer downfalls, and extra floods within the monsoon. Jointly for South-East Asia, the projected changes in rainfall depend on the season. Throughout the Northern hemisphere winter (December-February) rainfall might reduce whereas rainfall is maybe going extend at the June – August quantity. However, not only the standard annual or seasonal downfall will change, there will be increased at intervals that vary of significant downfall events resulting in further floods. Moreover, the number of days with rainfall is maybe going decrease that consequently will increase possibilities of prolonged periods without precipitation leading to additional frequent and additional severe droughts (Ludwig et al., 2007). Figure 5.8 shows Wheat was later harvested in northern Egypt.

FIGURE 5.8   Wheat was later harvested in northern Egypt (Hamada, 2018).

The IPCC (2007) report specific that: observational proof from all continents shows that lots of natural systems are stricken by regional climate changes. Most of the determined changes are coupled with temperature changes. Unluckily, most of the studies and observations are from the developed world and there is an important lack of data from developing countries. Those studies offered that specialize in the developing world show that Sahel ecological zones have shifted as a result of a dry and warmer climate that has jointly caused a reduction in run-off (Van Duivenbooden et al., 2002; Gonzalez, 2001). In Southern Africa, an extended time of the dry season and additional unsure rainfall has reduced agricultural production and has forced us to adapt through change crops, diversifying livelihoods, and planting trees (IPCC, 2007). In Africa, lower lake levels have been observed in the Republic of Zimbabwe, Republic of Zambia, and Malawi.

5.11 THE WORSE THE IMPACTS AND THE HIGHER THE FUTURE COSTS WILL BE

Without serious mitigation and adaptation, climate change is maybe going to possess a significant impact on developing countries and thus the poor are doubtless to suffer most. The later serious actions are taken, the additional serious the impacts and thus the upper the long-run prices are going be. Taking action presently on temperature change mitigation and adaptation will significantly reduce future injury. Therefore, on cut back future temperature change, efforts to cut back the emissions of greenhouse gases at intervals the developed world must be accumulated. The EU must still stimulate this within their member countries and improve efforts to stimulate the USA to join the Kyoto agreement and commit itself to future targets. If the developed countries do further to chop back their emissions, quickly developing countries are much more in all probability to joint mitigation efforts. Getting developing countries to commit to mitigation is extremely important as reducing future emissions of India and China is very important for fastness future temperature change. The EU ought to have fully completely different focuses for varied countries in terms of temperature change adaptation and mitigation, for quickly developing countries like China and India must target mitigation of greenhouse gas emissions and for the least developed countries (LDCs) there must be a spotlight on adaptation (Ludwig et al., 2007).

Greenhouse gas emissions from developing countries are increasing quickly. Reducing the enlargement of emissions from these countries is in the interest of the EU as a result of it will abate climate change. In developing countries, masses are drained terms of mitigation whereas not retardation down the economy; significantly in terms of fast energy efficiency, reducing deforestation and rising efficiencies in agriculture. It's in addition very important that developing countries are live aroused to choose on a property, low emission development pathway. It's always copious easier and value-effective if the choice for the extra property, low emission technologies is made early at intervals the tactic. Currently, the EU is stimulating the mitigation and transfer of contemporary technologies through the clean development mechanism (CDM). Though it's still unclear what the mitigation potential of the CDM is, significantly in the Republic of India the investment in CDM comes is very important. However, to primarily stimulate mitigation in developing countries, the EU must take how a wider approach in stimulating property development in these countries. Clearly, it ought to be left to the private countries on but they develop but significantly, in terms of production of merchandise foreign at the EU; there are live potentialities to stimulate cleaner production. For example by developing labeling systems or import duties relying on greenhouse emission and/or on but property product is a measure made. These mechanisms mustn't be used as a replacement instrument for policy but must stimulate further environmental-friendly production. Reducing GHG emissions must be integrated into consecutive spherical of trade negations and thus the global organization must conjointly acknowledge the role of interchange inflicting and preventing dangerous temperature change. Currently, the world market is stimulating the assembly of merchandise at very cheap worth whereas not taking the climate into thought. By swing a value on gas emissions, there will be a stimulation to supply products where it's through with very cheap energy input (for production and transport) (Ludwig et al., 2007).

Also, at the smallest amount, developed countries there are selections for mitigation but they need to not think about the energy or transport sector however on agriculture and biology. Initial of all, new mechanisms must be developed in such a way that the protection of forest is paid through well-constructed carbon markets. The EU must actively support post-Kyoto mitigation selections in reduced deforestation and/or forest conservation, but as long because it does not provide incentives for Annex-I countries to understand most of their commitments abroad (Ludwig et al., 2007). This may be achieved by mixtures of higher commitments, imposing limits or by turning out with associate freelance mechanism, committing developed countries to every national emission reductions and supporting reductions abroad. In agriculture, the EU must quest for win-win things where every productivity and/ or efficiency are inflated whereas not compromising the atmosphere, whereas at a similar time reducing emissions. Significantly the tiniest quantity developed countries must get facilitate from the EU for climate change adaptation.

Currently, most funds out there are through UNFCCC methodology. There are many complaints from developing countries that it's very exhausting to urge funding for adaptation comes through the GEF. The recently started GCCA might provide a probability to work on adaptation outside the sphere of negotiations. In addition, there is an associate increasing accord that the funds presently out there are not enough to support developing countries to cater to the impacts of climate. What seems forgotten at the discussion concerning funding for adaptation is that nearly all adaptation is extraordinarily rather like sensible development practices. The shut links between adaptation and property development makes it much more sense to thought adaptation into property development and built higher links between the distinction and development funds (Ludwig et al., 2007).

The key reason that lots of developing countries are in danger of temperature change is also an absence of accommodative capability. General accommodative capability and resilience are maybe going extend with development. However, to want specific adaptation measures an exact cognition is crucial. The cognition on climate change is usually very restricted in developing countries. There is a lack of data, studies, and trained personnel. The EU must actively support increasing (scientific) information on climate amendment impacts and changes in developing countries and improve capability building on temperature change adaptation. Still, lots of capability at the smallest amount developed countries is that concentrate on mitigation (e.g., CDM training) whereas these countries would profit much more from capability building related to adaptation (Ludwig et al., 2007).

5.12   TO PROMOTE ADAPTATION IN DEVELOPING COUNTRIES

To promote adaptation in developing countries, the EU must act at each world and European level (Ludwig et al., 2007):

- Within the context of the UNFCCC, the EU will still advance the problem of adaptation, and promote the mix of adaptation into national development plans (e.g., through the National Adaptation Programs of Action (NAPA) and thus the 5-Year Work Program on Adaptation that's recently adopted in Nairobi). EU leadership is going to be required to help confirm the accessibility of ample financial and technical resources, along with the distinction Fund beneath the Kyoto Protocol (KP), the worldwide atmosphere Facility and bilateral channels, to implement NAPAs and similar ways.

- The 2004 EU Action arrange on Climate change and Development already includes support ways for adaptation in developing countries which will, as an example, be supported beneath the Environment and Natural Resources Thematic program and via geographical funds at the country and regional level. The inclusion of adaptation measures in geographical programming will need to be compelled to be sturdy. The consecutive occasion for this can be often the mid-term review of country and regional ways in 2010. The continued mid-term analysis of the climate change arranges provides the first probability for a review of the arrangements at the light-weight of quick climate change.

- The Commission is examining how to push and increased dialogue and cooperation between the EU and developing countries on climate change, through the building of a world climate change Alliance. The Commission has earmarked a whole of €50 million over the period 2007–2010 for dialogue activities, and to support developing countries through targeted mitigation and adaptation measures. Actions might embrace providing follow-up to the NAPA through concrete pilot comes significantly referring to the integration of adaptation activities in key sectorial policies. Moreover, the forthcoming EU strategy on Disaster Risk Reduction will build a bridge between adaptation and disaster response. Figure 5.9 shows a successful framework must directly involve assistance for adaptation in developing countries.

FIGURE 5.9   A successful framework must directly involve assistance for adaptation in developing countries (Hamada 2018).

5.13   KEY RESULTS OF THE GLOBAL CLIMATE RISK

The Global Climate Risk Index 1995–2014 relies on the medium values over a 20-year quantity. However, the list of states featured at the Bottom 10 is split into two groups: those that only have a high ranking to exceptional catastrophes and those that have unceasingly extreme events. Countries falling into the previous category embraces Myanmar, where Cyclone

Naris in 2005 caused quite 95% of the injury and fatalities at intervals the past 20 years, and the Republic of Honduras, where quite 80% of the injury in every category was caused by cyclone Mitch in 1998.

The most recent addition to the current cluster is Thailand, wherever the floods of 2011 accounted for 87% of the general injury. With new super-latives like cyclone Patricia in Oct month 2015 being that the strongest land-falling pacific cyclones on record, it seems to be merely a matter of it slow until a consecutive exceptional catastrophe happens. Cyclone Pam, that severely hit Vanuatu in March 2015, yet again showed the vulnera-bility of LDCs and Small Island Developing States (SIDS) to climate risks. Countries rather like the Philippines, West Pakistan, and the Republic of India are vulnerable by extreme weather events every annum and keep at the Bottom 10. As a country that is full of 8–9 typhoons per year and thus the victim of outstanding catastrophes, significantly cyclone Haiyan in 2013, the Philippines suggests that a replacement and distinctive classifica-tion of states that match every mold is additionally rising. Similarly, the appearance of some European countries among the underside 30 countries can to an oversized part be attributed to the extraordinary vary of fatali-ties to the 2003 heat wave, throughout that quite 70,000 of us died across Europe. Though a number of these countries are typically hit by extreme events, the relative economic losses and thus the fatalities are typically rela-tively minor compared to the countries' populations and economic power. However, the Republic of Bosnia and Herzegovina lost nearly one-tenth of its GDP because of the 2014 flooding (Kreft et al., 2016).

5.14 AFRICA HIT HARDEST BY GLOBAL WARMING DESPITE ITS LOW GREENHOUSE GAS EMISSIONS

The continuous increase in average world temperatures is very caused by an increase in greenhouse gases emitted through human activities. Warming threatens dangerous consequences like drought, diseases, floods, and lost ecosystems. Although it's notable that the richest nations are inflicting the greater warming, poor countries suffer the foremost. Africa is very in danger of the implications of global warming as a result of it over-whelmingly depends on rain-fed agriculture, many African countries are making efforts to adapt to climate change; however, national efforts ought to be complemented by international cooperation to realize the desired results. To tackle the challenge shown by climate change the Convention on climate change set associated overall framework for intergovernmental efforts. In 1997, several parties by signing the KP agreed to specific targets for reducing their emissions of greenhouse. As part of the effort to handle temperature change issues versatile mechanisms, that embrace CDM, were introduced; however, the CDM strategy, that was half designed to produce non-Annex I countries new funding for property development, has not helped Africa. Figure 5.10 shows the Mediterranean Sea rises and hits the Alexandria coast in Egypt.

FIGURE 5.10   Mediterranean Sea rises and hits Alexandria coast, Egypt (Hamada 2018).

It is unfair to expect the non-Annex I countries to make emissions reduction as warming is caused by industrialize, besides, developing countries need to be compelled to extend their own emissions to satisfy the wants of development. Developed nations mustn't only produce efforts to lower their gas emissions however also help developing countries by mobilizing financial resources, technical facilitate and capability building programs. It's significantly powerful for Africa to realize climate-friendly sustainable development on its own land and thus developed nations and international non-government organizations must help Africa deal with the impacts of worldwide warming.

5.15 CONCLUSION

This chapter provides a quick review of background of climatic influences on crop yields and net farm incomes. Where, weather change presents pretty complex and numerous challenges for growing countries, mainly for low earnings countries along with malnutrition, hunger, forced migration, poverty and seasonal unemployment. Agriculture is an especially important sector in Africa, contributing toward livelihoods and economies across the continent. On average, agriculture in Sub-Saharan Africa contributes 15% of the overall GDP. Africa's geography makes it specifically prone to climate change, and 70% of the populace depend upon rain-fed agriculture for their livelihoods. Smallholder farms account for 80% of cultivated lands in sub-Saharan Africa. The Intergovernmental Panel on Climate Change (IPCC) projected that climate variability and change would critically compromise agricultural production and access to food. Cropping systems, farm animals and fisheries may be at greater chance of pest and sicknesses as an end result of future climate change. Research software on Climate Change, Agriculture and Food Security (CCAFS) have identified that crop pests already account for about 1/6 of farm productivity losses. Similarly, weather change will accelerate the superiority of pests and sicknesses and increase the occurrence of highly impactful events. The impacts of climate change on agricultural production in Africa will have serious implications for food security and livelihoods. Between 2014 and 2018, Africa had the highest levels of food insecurity within the world.

KEYWORDS

- climate change
- crop yields
- Food and Agriculture Organization (FAO)
- Greenpeace
- Intergovernmental Panel on Climate Change (IPCC)
- international temperatures

REFERENCES

Ellis, S., (2008). *The Changing Climate for Food and Agriculture Institute for Agriculture and Trade Policy Minneapolis.* Minnesota, ©2008 IATP. All rights reserved.

Fulco Ludwig, Catharien Terwisscha van Scheltinga, Jan Verhagen, Bart Kruijt, Ekko van Ierland, Rob Dellink, Karianne de Bruin, Kelly de Bruin, & Pavel Kabat (2007). Climate change impacts on Developing Countries - EU Accountability, Wageningen University and Research Centre Droevendaalsesteeg 4, 6708 PB Wageningen, The Netherlands www.wur.nl Co-operative Programme on Water and Climate (CPWC) Westvest 7, 2611 AX Delft, The Netherlands www.waterandclimate.org.

Gonzalez, P., (2001). Desertification and a shift of forest species in the West African Sahel. *Climate Research, 17,* 217–228.

Hamada, Y. M. (2018). Special pictures album, Took by Youssef M. Hamada, Egypt, 2018.

IISD and EARG, (1997). *International Institute for Sustainable Development and Environmental Adaptation Research Group.* Institute for environmental studies, University of Toronto. Agriculture and climate change, a prairie perspective.

IPCC, (2007). *Climate Change 2007, Fourth Assessment Report.* Cambridge University Press, Cambridge, UK. www.ipcc.ch (accessed on 4 January 2020).

Ralph, J. C., & Paul, N., (2016). *Climate Change Evidence and Causes.* An overview from the Royal Society and the US National Academy of Sciences.

Sönke, K., David, E., Lukas, D., & Livia, F., (2016). *Global Climate Risk Index.* German watch, www.germanwatch.org (accessed on 4 January 2020).

UNFCCC, (2007). *Climate Change: Impacts.* Vulnerabilities and adaptation in developing countries, the United Nations Framework Convention on Climate Change. www.unfccc. int (accessed on 4 January 2020).

Van, D. N., Abdoussalam, S., & Ben, M. A., (2002). Impact of climate change on agricultural production in the Sahel: Part 2. *Case Study for Groundnut and Cowpea in Niger Climatic Change, 54,* 349–368.

Wheaton, E. E., (1994). *Impacts of a Variable and Changing Climate on the Canadian Prairie Provinces: A Preliminary Integration and Annotated Bibliography.* SRC Publication, No. E-2900.7.E.93. Saskatchewan Research Council, Saskatoon.

CHAPTER 6

Climate Change and Food Security: An Overview

ABSTRACT

Climate change will have a control on all four scopes of food security: food obtainability, food convenience, food depletion, and food systems constancy. It will have an impression on human fitness, existing assets, food-producing, and rotation canals, additionally as dynamical buying rules and market innings. Its effects are either short term, future from lots of frequent and lots of intense risky weather events, and long-standing, caused by precipitation patterns, people those that are already prone and food diffident are probable to be the primary affected. Agriculture-based living structures that are already weak to food insecurity face the immediate danger of deficiency crop production, new patterns of diseases and pests, deficiency of applicable seeds and planting material, and loss of cattle. People living on the coastlines and valleys and in mountains, drylands and therefore the Arctic are most at hazard. As an associate indirect consequence, low-income people all over, however largely in city areas, are at danger of food uncertainty because of loss of possessions and deficiency of adequate assurance coverage. This might in addition cause shifting susceptibilities in every developing and developed country. Food systems additionally are affected through possible internal and international relocation, resource-based mostly fights and civil strife triggered by climate change and its effects (FAO, 2008). This chapter displays the problem of climate change and food security (CCFS) as the views of most of the specialists in CCFS.

6.1 BACKGROUND

Food security is that the end result of food system processes all along with the nourishment series. World climate change will have a control on food security bit through its impacts on all components of worldwide, national, and domestic nourishment systems. World climate change is real, and its initial impacts are already being felt. It will initially have a control on the people and food systems that are already vulnerable, but over time the geographic distribution of risk and vulnerability is maybe visiting shift. Certain support teams would wish immediate support, but everyone seems to be in peril. Risk exists once there is uncertainty regarding the long-term outcomes of current processes or regarding the prevalence of future events. Adaptation is regarding reducing and responding to the one risks world climate change poses to people's lives and livelihoods. Reducing uncertainty by up the knowledge domain and making innovative schemes for insuring against world climate change hazards will ever be necessary for effective adaptation. Adjective management is a really valuable tool for making ways that reply to the distinctive risks thereto utterly completely different scheme and resource groups are exposed. Strengthening resilience involves adopting practices that enable vulnerable people to defend existing resource systems, diversify their sources of economic gain, modification their resource ways or migrate, if this could be the foremost effective chance. Dynamical consumption patterns and food preparation practices may even be enough to defend food security in many circumstances (Figure 6.1).

Each process and voluntary selections influence individual choice regarding what food to eat and also to be sure of physiological condition below a dynamical climate. Safeguarding food security among the face of world climate change, in addition, implies avoiding the disruptions or declines in the world and native food provides that may result from changes in temperature and rainfall precipitation regimes and the new patterns of pests and diseases (FAO, 2008).

Raised productivity from improved agricultural water management is crucial to creating certain world food provide and world food security. Property livestock farming management practices for adaptation and associated mitigation must even lean high priority. Conservation agriculture can produce an enormous distinction to the efficiency of water use, soil quality, the capability to resist extreme events, and carbon dioxide decrease. Promoting agro variety is particularly necessary for topical adaptation and resilience. Meeting the growing demand for energy may be a necessity for sustainable growth and development. Bioenergy is maybe going play associate more and additional necessary role, but its use should

FIGURE 6.1   Adapting to climate change for sustainable agribusiness (Hamada, 2018).

not undermine food security. Mitigating world climate change suggests that reducing gas emissions and sequestering carbon dioxide or storing it in the short term, major importance, creating development selections that can decrease risk in the future. Though the whole food system may be a provider of carbon dioxide emissions, primary production is out and away the foremost necessary half. Incentives are needed to influence crop and livestock producers, agro-industries and scheme managers to adopt good practices for mitigating world climate change (FAO, 2008).

In the food and agriculture sector, adaptation, and mitigation over-poweringly go hand in hand, so adopting an associate integrated strategic approach represents the foremost effective approach forward. Several funds at intervals the international organization system finance specific activities geared toward reducing gas emissions and increasing resilience to the negative impacts of world climate change. As a results of many mitigation actions that may have high payoffs, in addition, represent good selections for adaptation at intervals the food and agriculture sectors of low-income developing countries, it's visiting be getable to urge more resources from bilateral and multilateral aid agencies, that became additional affected by finance development resources in responses to world climate change (FAO, 2008).

6.2   SCOPE OF THE CHAPTER

Agriculture, biology, and fisheries will not entirely be influenced by world climate change, but in addition, contribute thereto through emitting greenhouse gases. They furthermore hold part of the remedy, however; they're going to contribute to world modification global climate change temperature change mitigation through reducing gas emissions by change agricultural practices. At constant time, it is necessary to strengthen the resilience of rural individuals and to help them with address this additional threat to food security. Notably among the agriculture sector, world climate change adaptation will go hand-in-hand with mitigation. World climate change adaptation and mitigation measures have to be compelled to be integrated into the event approaches and agenda. Until recently, most assessments of the impact of world temperature change on the food and agriculture sector have targeted on the implications for production and world provider of food, with less thought of different components of the organic phenomenon. This chapter takes a broader scan and explores the multiple effects that warming and world climate change could wear food systems and food security. It, in addition, suggests ways for mitigating and adapting to world temperature change in several key policy domains of importance for food security (Killmann, 2008). This chapter provides background data on the interrelationship between world climate change and food security (CCFS), and ways in which to handle the new threat. It additionally shows the opportunities for the agriculture sector to adapt; also as describing, however, it will contribute to mitigating the climate challenge.

6.3 FOOD SECURITY

In 2007, at the 33rd session of the Committee on World Food Security, Food, and Agriculture Organization of the world organization issued an announcement to affirm its vision of a food-secure world: FAO's vision of a world without hunger is one in the majority are ready themselves to urge the food they require for lively and healthy life, and wherever social safety nets ensure that those lack resources still get enough to eat (FAO, 2007c). This vision has its roots within the definition of food security adopted at the world food summit (WFS) in Nov 1996: Food security exists once all individuals the least bit have physical or economic access to sufficient safe and wholesome food to fulfill their dietary desires and food preferences for a vigorous and healthy life (FAO, 1996). WFS, the Inter-Agency unit that established the food insecurity and vulnerability data and mapping system (FIVIMS) careful an abstract framework that gave operational intending to this definition.

Food and Agriculture Organization (FAO) of the United Nations (UN) reaffirmed this scan in its initial written assessment of the implications of world climate change for food security, contained in its 2015 to 2030 projections for world agriculture. FAO of the UN stressed that food security depends lots of on socio-economic conditions than on agro condition zones and on access to food rather than the assembly or physical convenience of food. It expressed that, to judge the potential impacts of world climate change on food security, it is not enough to assess the impacts on domestic production in food-insecure countries. In addition, one has to (i) assess world climate change impacts on exchange earnings; (ii) verify the pliability of food surplus countries to increase their industrial exports or food aid; and (iii) analyses how the incomes of the poor are laid low with climate change (FAO, 2003b).

6.4 FOOD SYSTEM

Definitions of food security confirm the outcomes of food security and are helpful for formulating policies and preferring actions, but the processes that cause desired outcomes additionally matter. Most current definitions of food security, so embrace references processes additionally as outcomes. Recent work describing the functioning of food systems has helped to point every desired food security goals and what must happen to bring these regarding. Between 1999 and 2003, a series of specialized consultations, convened by the Global environmental modification and food systems (GECAFS) project with FAO's participation, developed a version of the FIVIMS framework that more clarifies, however, a range of processes on an organic phenomenon must occur so as to bring forth food security. Taken on, these processes represent the food system, and so the performance of the food system confirms whether or not food security is achieved (FAO, 2008).

6.5   FOOD CHAIN

The total of all the processes during a food system is every now and then noted as a food sequence, and sometimes given catchy slogans like from the plow to plate or from farm to fork. The foremost abstract distinction between a food system and a food chain is that the system is holistic, comprising a bunch of at the identical time interacting processes, whereas the chain is linear, containing a sequence of activities that require to occur for individuals to get food. The concept of the food system is useful for scientists to work cause and impact relationships and feedback loops, and is extremely necessary for the technical analyses that underpin policy recommendations. However, once humans action the findings of such investigations it's usually easier to use the concept of the food chain (FAO, 2008).

The section on Food security and global climate change: an abstract framework presents a simplified description of the dynamics of poten-tial climate change impacts and feedback loops in an exceedingly very holistic food system. The implications are mentioned linearly, however, by projected changes for every of five of the foremost necessary climate variables for food systems, and at the potential impacts of each of these changes on each food system technique. A food system contains multiple food chains operating at the global, national, and native levels. Varieties of those chains are thus short and not thus complicated; whereas others circle the global in the associate labyrinthine net of interconnecting processes and links. One straight forward chain, that's important for food security in many households active in rain-fed agriculture, begins with a staple cereal crop made during a farmer's field, moves with the harvested grain through a locality mill and back to the farmer's home as luggage of flour, and finishes among the cookery pot and on the house members' plates. This same house altogether chance, in addition, participates in an exceedingly lots of sophisticated food chain to bring salt, that's domestically on the market in mere a few of the places, but is utilized worldwide as a preservative and seasoning. Part of the meager cash gain of even the poorest farming households is sometimes lost sight of to shop for salt from passing traders or native stalls. A household's food system contains all the food chains it participates in to satisfy its consumption requirements and dietary preferences, and each one the interactions and feedback loops that connect the assorted elements of these chains. This instance shows that it isn't potential that a house can do food security whereas not some cash expenditure. All households would like sources of livelihood that provide them spare buying power to shop for the food that they need but cannot or do not manufacture for their own consumption (FAO, 2008).

Climate may be a notably necessary driver of food system performance at the farm end of the food chain, poignant the quantities and types of food created and so the adequacy of production-related gain. Extreme weather events can damage or destroy transport and distribution infrastructure and have a control on different non-agricultural elements of the food system adversely. However, the impacts of world temperature change are in all probability to trigger adjective responses that influence the environmental and socio-economic drivers of food system performance in positive additionally as negative ways in which. This chapter is concerned with the projected balance of those numerous impacts on food system performance and food security outcomes at the native and world levels.

6.6 AGRICULTURE AND CLIMATE CHANGE

Agriculture contributes to climate greenhouse gas emission and intensely influenced of alteration in climate factors. In intensive farming, we have an inclination to expect high greenhouse gases emission because of using a high amount of inputs and chemicals, due to these changes of act natural various and world temperature change impacts varies consequently in many parts of the world.

Vulnerability to world temperature change depends not entirely on material and biological responses but in addition to socio-economic characteristics, as low-revenue population notably people who cultivate crops below rain-fed and non-irrigated agriculture systems in drylands, arid, and semi-arid areas extraordinarily due to world climate change (Grasty, 1999).

6.6.1   CLIMATE VARIABLES AND PRODUCTIVITY

According to suggestion of fourth assessment, report of IPCC (2007a) the gross impacts of higher temperatures on crop responses at the plot level, whereas not considering changes among the frequency of most events, moderate warming may increase crop and pasture productivity in temperate regions, whereas it's going to reduce productivity in tropical and semi-arid regions. Modeling studies indicate little helpful impact in temperate that corresponds to native mean temperature, can increase from 1–3°C, with an association of the increase in carbon dioxide gas and downfall rainfall changes. In a disparity, models show that orbital regions show a negative yield impacts for major crops with a moderate rise in temperature (1–2°C), but any warming projected all regions among the end of the twenty-first century ends up in the increased on negative impacts (Tubiello et al., 2007).

Sequent to impacts of world climate change, agricultural productivity is directly affected in the developed and developing world (Alexandrov and Hoogenboom, 2000). Climate plays a major role in determinative the yield level by increasing or reducing in world's perspective from temperate to tropics. Many experiments show that carbon dioxide gas may be a limiting issue, throughout that one higher concentration of carbon dioxide gas enhances 25 natural action and crop growth, modifying water and nutrient cycles (Tubiello et al., 2007), these responses found to hold even for plants matured beneath completely different stress conditions.

Despite related to the generalization of the law of `limiting factors' once different environmental factors like water shortage, less light, shortage or surplus of minerals, very high or very low limit yield, then carbon dioxide gas concentration will have little or no impact. However, in stressful environments, the relative natural action response of plants to carbon dioxide gas richness is actually increased. For several crops, that growing below elevated carbon dioxide gas conditions every quality and total yield shows improved (more ear of plant per m2). The raised value carbon dioxide gas induce and makes an increase among the grain weight and in line with the observation, it had been larger below average phosphorus treatment compared to higher phosphorus level. This influence of carbon dioxide gas and phosphorus provide was attributed to increase among the variability of cells at reproductive structure, that's that the result of the raised rate of process throughout grain development or by a larger amount of grain filling throughout ripening stage (Uprety et al., 2010). However, it's in addition been shown that top carbon dioxide gas concentrations may have negative effects on the grain quality from wheat in terms of crude protein content (Pleijel and Uddling, 2011), it alters wheat grain lipids and doubled the full of mitochondrion in wheat leaves, lower seed element concentration and reduces grain and flower protein (Qaderi, 2009).

6.6.2 DIRECT EFFECTS OF CLIMATE CHANGE ON FOOD CROPS

Food production is negatively or fully affected following variation in weather patterns (short winter, long summer, earlier spring) and different extreme weather events like drought (change in the amount and temporal range of precipitation), flooding, etc. In addition, non-legal deforestation will cause a reduction in crop production, due to its impact on environmental services like crop fertilization, genetic resources, clean air, and water offer, soil fertility, and erosion, additionally as pests and microorganism management (Cerri et al., 2007).

6.6.3 INDIRECT EFFECT OF TEMPERATURE

Several studies have shown that soil warming can have an impact on accessibility of nutrient, increase soil N mineralization and nitrate action, organic matter decomposition, and a little temperature increase can manufacture an enormous sweetening of activities. A raise in N mineralization in the soil are foreseen beneath favorable wetness conditions and substrate availableness, in the main in those ecosystems where the temperature may be a limiting issue, that leads to increased NPP (net primary production), increased N demand and ultimately to decrease N availability among the soil. Increasing temperature too can speed up the releasing of nutrients interlock up in organic soil fraction and minerals, whereas decreasing soil wetness may limit this technique. A preferable robust rate of weathering of nutrient created rocks usually leads to higher base saturation of the soil and maintains higher soil pH, every characteristic favorable to plant growth. Whereas elevated carbon dioxide gas not thought to possess a right away impact on weathering (Lukac et al., 2010).

Plant species distribution is restricted not entirely by their absolute limits of survival, but in addition through competition within species, in which species of acclimatize and grow higher in an exceedingly given climate. Among the context of population extinction, it is necessary to ponder the results throughout climate events. Temporal variability in surroundings, as a rule, believed to increase the probability of population extinction, notably if environmental variability can increase due to world climate change. Some proof suggests world climate change already drives the extinction of rear edge plant populations leading to a distribution with an edge. Some desert trees like Aloe dichotomy in southern Africa the sting of vary making populations showed negative demographic rates, and powerfully positive rates determined at the forefront of the vary making increment rate sensitive and use a full indicator of early modification in vary (Thuiller et al., 2008).

6.6.4   IMPACTS OF TEMPERATURE × [CO₂] INTERACTION ON PLANT PROCESSES

There are many processes in plant growth, filled with the interaction of every raised temperature and carbon dioxide gas, in processes that verify carbon balance among the shorter term, from the long-standing scales of development and growth, that on cause accumulation of biomass and yield. The two main reasons to expect additional and additional increasing oxide gas responsiveness of plant carbon balance at higher temperatures are: 1) the diminished quantitative relation of chemical change to respiration and 2) the diminished quantitative relation of gross chemical change to dark respiration within the hotter conditions (Morison and Lawlor, 1999).

The impact of elevated carbon dioxide gas on photosynthetic process change reactions are lots of pronounced in heat, e.g., around 20°C than at 10°C. Some predictions indicate that future increases in temperature may increase root mortality lots of in N-rich soils in temperate forests than in N poor soils in boreal forest areas with necessary implications for the cycling sport between plant and soil (Lukac et al., 2010). Some (unpublished) studies found that changes in activation state and constant chemical change occur due to each of the carbon dioxide gas and temperature, and there is an interaction, that affected the chemical process rate demonstrating the underlying quality of the chemical process regulation mechanisms (Morison and Lawlor, 1999). To sum up, the environmental modification includes a sway on the rate of growth of individual trees and have an additive result on utterly completely different interactions and processes in the forest and has the potential to change the quantities of living materials among the forest scheme as a complete (Lukac et al., 2010).

Temperature is one in each of the decisive factors in forming the impact on growth and productivity by quick bud burst (BB), flowering, and stems elongation throughout spring then and extend the growing season, and it's one in every of the most factors dominant species distribution. For instance, the expected warming of 2–6°C by 2100 in north temperate forest regions could have substantial impacts on growth and species composition (Gunderson et al., 2012). Increasing temperatures, for the most part, go along with elevated carbon dioxide gas, vapors pressure deficit (VPD) and drought, modification in temperature will act with different factors to form associate effect; for e.g., lots of other processes are influenced by rising temperature. The long-term responses of world climate change below higher carbon dioxide gas concentration, temperature, and precipitation may disagree from short effects because of the feedbacks involving nutrient cycling (Chen et al., 1996). Tree seedlings exposed in elevated $[CO_2]$ over amount of less one year resulted in raised rate of photosynthesis, shrunken in respiration and increased growth, with little increase in leaf space and tiny variation in carbon allocation.

Exposure of woody species in elevated carbon dioxide gas over a long time period may result in higher rates of photosynthesis, but net carbon accumulation may not mostly increase if carbon dioxide gas unhitches from soil respiration increases' (Luxmoore et al., 1993). Environmental shift affects the plant diseases and insect pests in each of the manifestation and the infestation, introduced of the new species. Following these changes, the sort of diseases, pests, and weeds, preventing actions needed to chop back the results on human health and ecosystems (Roos et al., 2010). Utterly completely different chemical, biological, and physical processes in earth systems would wish varied temperature ranges; generally moderate and optimum temperature (for each process) is essential for classical activities at intervals the systems, an exact rise or lower from moderate temperature will have an impact on many activities inside the processes.

6.7   CLIMATE AND ITS MEASUREMENT

Climate refers to the characteristic conditions of the earth's lower surface-atmosphere at an exhaustive location; weather refers to the day-to-day fluctuations in these conditions at the same here location. The variables that are live typically used by meteorologists to measure daily weather phenomena are air temperature, precipitation (rain, sleet, snow, and hail), gas pressure and humidity, wind, and sunshine and cloudiness. Once these weather phenomena are measured systematically at a selected location over a few years, a record of observations is accumulated from that averages, ranges, maximums, and minimums for each variable are computed, along with the frequency and length of lots of maximum events (FAO, 2008).

The World meteoric Organization (WMO) requires the calculation of averages for consecutive periods of 30 years, with the foremost recent being from 1961 to 1990. Such a time is long enough to eliminate year-to-year variations. The averages are utilized within the study of world climate change, and as a base, therewith current conditions are compared (UK Met workplace Online). Climate is delineating at utterly completely different scales. World climate is that the typical temperature of the earth's surface and so the atmosphere connected with it, and is measured by analyzing thousands of temperature records collected from stations everyplace the world, every ashore and toward land. Most current projections of world temperature change attribute with world climate, but climate will even be delineated at different scales, based mostly records for weather variables collected from stations among the zones concerned. Zonal climates embrace the subsequent (FAO, 2008):

• Corners climates are temperature regimes determined by the position north or south of the equator. They contain the polar climate, temperate climate, sub-tropical climate, and tropical climate.
• Regional climates are patterns of weather that have an impact on an uneven geographical area which can be famous by special choices that distinguish them from different climate patterns. The key factors determinative regional climates are:

    i.   Variations in temperature caused by distance from the equator and seasonal changes among the angle of the sun's rays as the planet rotate;

ii. Planetary distribution of land and ocean areas; and

iii. The worldwide system of winds, attribute to the overall spreading, that arises as a result of the temperature distinction between the equators and poles. As models of regional climates are maritime climate, continental climate, monsoon climate, Mediterranean climate, coastal climate, and desert climate.

- Local climates have influence over very little geographical areas, of entirely a pair of kilometers or tens of kilometers across. They contain all land and ocean breezes, the orographic lifting of the air plenty and formation of clouds on the weatherboard of mountains, and the warmth effects of cities. Below certain conditions, native condition effects may predominate over lots of general pattern of regional or the angular distance climate. If the realm involved is extraordinarily little, as in an exceedingly flower bed or a shady grove, it's visiting be noted as a microclimate. Microclimates will even be created by artificial means, as in hothouses, exhibitions, or storage environments where temperature and humidity are controlled.

6.8 CLIMATE SYSTEM

The climate system is extremely sophisticated. Below the influence of the sun's radiation, it determines the earth's climate (WMO, 1992) and consists of:

- **The Atmosphere:** Volatilized state on the earth's surface;
- **The Hydrosphere:** Liquid water on or below the earth's surface;
- **The Cryosphere:** Snow and ice on or below the earth's surface;
- **The Lithosphere:** Earth's surface (rock, soil, and sediment);
- **The Biosphere:** Earth's plants and animal life, embrace humans.

Although climate intrinsically relates entirely to the variable states of the earth's atmosphere, the alternative elements of the climate system even have necessary roles in forming climate, through their interactions with the atmosphere. The Global Climate Observing System (GCOS) has developed an inventory of variables essential for monitor changes among the climate system. The list includes atmospherically, oceanic and terrestrial phenomena, and covers all the spheres of the climate system. GCOS was established by the Intergovernmental Oceanographic Commission (IOC) of the United Nations Educational, Scientific, and Cultural Organization (UNESCO), the international organization surroundings Program (UNEP) and so the International Council for Science (ICSU) in 1992 to form certain that the observations and data required handling climate-related problems are obtained and created accessible to all or any potential users. GCOS and its partners provide necessary and continuous support to the international organization Framework Convention on world temperature change (UNFCCC), the world Climate analysis Program (WCRP) and so the Intergovernmental Panel on world temperature changes (IPCC). The report system on essential climate variables provides data to (GCOS on-line a):

o   Characterize the state of the global climate system and its variability;
o   Observe the forcing of the climate system, by each of natural and anthropogenesis causes;
o   Support attributions of global climate change causes;
o   Support predictions of worldwide climate change;
o   Modify projection of worldwide world temperature change data to the regional and native scales;
o   Modify characterization of most events that are necessary in impact assessment and adaptation, and to the assessment of risk and vulnerability.

6.9   CLIMATE VARIABILITY AND CLIMATE CHANGE

There is no international agreement definition of the term climate change. World climate change can refer to: (i) long-term changes in average climate (WMO usage); (ii) all changes among the climate system, along with the drivers of modification, the changes themselves and their effects (GCOS usage); or (iii) entirely human-induced changes among the climate system (UNFCCC usage). There's, in addition, no agreement on the way to stipulate the term climate variability. Climate has been in an exceedingly very constant state of modification throughout the earth's 4.5 billion-year history, but most of these changes occur on astronomical or geological timescales, and is too slow to be determined on a humanitarian being scale. Natural climate variation on these scales is sometimes noted as climate variability, as distinct from human-induced world temperature change. UNFCCC has adopted this usage (UNFCCC, 1992). For meteorologists and climatologists, however, climate variability refers entirely to the year-to-year variations of atmospherically conditions around a mean state (WMO, 1992). To assess world temperature change and food security, the Food and Agriculture Organization of the UN prefers to use a comprehensive definition of world temperature change that encompasses changes in long-term averages for all the essential climate variables. For many of these variables, however, the experimental record is simply too short to clarify whether or not recent changes represent true shifts in long-term suggests (climate change), or are simply anomalies around a stable mean (climate variability).

6.10 EFFECTS OF GLOBAL WARMING ON THE CLIMATE SYSTEM

Global warming is the immediate consequence of increased gas emissions with no countervailing that can increase in carbon storage on earth. This part is disquiet degrading in the main with the projected effects of this episode of human-induced warming on the climate system, presently, and for several decades, as these are the results which can every cause more stresses and build new opportunities for food systems, with resultant implications for food security (Killmann, 2008).

6.11 ACCLIMATIZATION, ADAPTATION, AND MITIGATION

Acclimatization is truly adaptation that happens impromptu through autonomous efforts. Adaptation to world temperature change involves deliberate changes in natural or human systems and behaviors to chop back the risks to people's lives and livelihoods. Mitigation of world temperature change involves actions to chop back gas emissions and impound or store carbon among the short term, and development selections which can cause low emissions among the future (Killmann, 2008).

Adjustment may be a forceful and effective adaptation strategy. In straightforward expressions, meaning that getting accustomed to world temperature change and learning to live well with it. All living organisms, along with humans, adapt, and develop in response to changes in climate and surroundings. Some variations may even be biological—as an example, human physiology may become lots of warmth tolerant as world temperatures rise—but many are in all probability to involve changes in perceptions and mental attitudes that reinforce new, lots of tailored responses to extreme events (Killmann, 2008). Figure 6.2 shows acclimatization, adaptation, and mitigation in Upper Egypt.

FIGURE 6.2   Acclimatization, adaptation, and mitigation in Upper Egypt, Egypt (Hamada 2018).

6.12   AGRICULTURE, CLIMATE, AND FOOD SECURITY

Agriculture is extremely necessary for food security in two ways: it produces the food people eat; and (perhaps the additional important) it provides the most provider of resource for 36% of the world's total workers. Among the heavily dwelled countries of Asia and thus the Pacific, this share ranges from 40 to 50%, and in sub-Saharan Africa (SSA), two-thirds of the operational population still produce their living from agriculture (ILO, 2007).

If agricultural production among the low-income developing countries of Asia and Africa is adversely stuffed with world climate change, the livelihoods of monumental numbers of the agricultural poor are place in peril and their vulnerability to food insecurity increased. Agriculture, agroforestry, and fisheries are all sensitive to climate. Their production processes are so in all probability to be influenced by world climate change. In general, impacts are expected to be positive in temperate regions and negative in tropical ones, but there is still unsteadily regarding but projected changes will play out at the native level, and potential impacts may even be altered by the adoption of risk management measures and adaptation by ways strengthen state and resilience (Killmann, 2008).

The food security implications of changes in agricultural production patterns and performance are two kinds:

- Impacts on the assembly of food will have a control on food provided at the global and local levels. Globally, higher yields in temperate regions could offset lower yields in tropical regions. However, in many low-income countries with the restricted financial capability to trade and high dependence on their own production to lid food requirements, it's going to not be getable to offset declines in native offerings while not increasing reliance on food aid.
- Impacts on every kind of agricultural production will have a control on livelihoods and access to food. Producer associations that are less able to modify with world climate change, just like the agricultural poor in developing countries, risk having their safety and welfare compromised. Different food system processes, like food method, distribution, acquisition, preparation, and consumption, are as necessary for food security as food and agricultural production are. Technological advances and so the event of long-distance promoting chains that move manufacture and prepackaged foods throughout the world at high speed and relatively low price have created overall food system performance manner less obsessive about climate than it had been 200 years ago.

However, as a result of the frequency and intensity of severe weather increase, there is a growing risk of storm damage to transit and distribution infrastructure, with resultant disruption of food provides chains. The rising price of energy and so the scale back to reduce fuel usage on the food chain have drove to a novel calculus—food miles, that must be unbroken as low as getable to slash back emissions. These factors could result in lots of native responsibility for food security, that must be thought of among the formulation of adaptation ways for folks that are presently vulnerable or who could become so at the sure future (Killmann, 2008).

6.13 FOOD SECURITY AND CLIMATE CHANGE: A CONCEPTUAL FRAMEWORK

Food systems exist in the region, along with all different manifestations of human commercial activity. The projected will increase in mean temperatures and downpour will not manifest through constant gradual changes, but will instead be practiced as increased frequency, length, and intensity of hot spells and precipitation events. Whereas the annual prevalence of hot days and most temperatures are expected in all parts of the globe, the mean world increase in precipitation is not expected to be uniformly distributed around the world. In general, it's projected that wet regions will become dry regions. For this analysis, an abstract framework on world temperature change and food security interactions were developed to target the variables shaping the food and climate systems. The world CCFS framework shows anywise climate change affects food security outcomes for the four components of food security, food availability, food accessibility, food utilization and food system stability-in direct and indirect ways (Killmann, 2008).

Climate change variables influence biophysical factors, like plant and animal growth, water cycles, multifariousness, and nutrient cycle, and also the ways throughout which these are managed agricultural practices and land use for food production. However, climate variables even have control over physical/human capital-like roads, storage, and promoting infrastructure, houses, productive assets, electricity grids, and human health-that obliquely changes the economic and socio-political factors that govern food access and utilization and would possibly threaten the stability of food systems. All of these impacts manifest themselves among the ways in that which food system activities are assigned. The framework illustrates but adjective changes to food system activities are needed all along the food chain to handle the impacts of world climate change (Killmann, 2008).

The world climate change variables thought of among the CCFS framework is:

- Increasing carbon dioxide gas concentrations among the atmosphere;
- Increasing extreme and minimum temperatures;
- Gradual changes in precipitation;
- Increase the frequency, length, and intensity of dry spells;
- Changes in the temporal arranging, duration, intensity, and geographic location of rain and snowfall;
- Increase among the frequency and intensity of storms and floods;
- Bigger seasonal weather variability and changes in the start/end of growing seasons.

Some of the various changes among the region that are expected to result from warming will occur among lots of distant future, as a consequence of changes in average atmospheric conditions. The fore likely things of world climate change indicate that may increase in weather variability and so the incidence of most weather events are notably necessary presently and among the immediate future.

6.14 VULNERABILITY TO CLIMATE CHANGE

Uncertainty and risk: Risk exists once there is uncertainty regarding the long-term outcomes of current processes or regarding the prevalence of future events. Lots of certain associate outcome is, the less risk, there is, as a result of certainty permits well-read selections and preparation to switch the impacts of risky processes or events. World climate change projections have a solid scientific basis, and there is growing certainty that extreme weather events are becoming to extend in frequency and intensity. This makes it extraordinarily in all probability that quality losses because of weather-related disasters will increase. Whether or not these losses involve productive assets, personal possessions or even loss of life, the livelihoods and food security standing of a lot of people in disaster-prone areas are adversely affected (Killmann, 2008).

6.15   FOOD SYSTEM VULNERABILITY

In overview, a food system became vulnerable where as one or lots of the four components of food security: food availability, food accessibility, food utilization and food system stability are unsure and insecure. Food availability is about by the physical quantities of food that are created, stored, processed, distributed and adjusted. Food and Agriculture Organization of the United Nations calculates national food balance sheets that embrace of those elements. Food accessibility is that the net quantity remaining once production, stocks and imports are summed and exports ablated for each item enclosed among the food record. Adequacy is assessed through comparison of availability with the estimated consumption demand for each food item. This approach takes into thought the importance of international trade and domestic production in reassuring that a country's food provide is sufficient. Constant approach will even be accustomed verify the adequacy of a household's food provide, with domestic markets enjoying the leveling role (Killmann, 2008).

High market prices for food are generally a reflection of inadequate availability; persistently high prices force poor people to chop back consumption below the minimum required for a healthy and active life, and can cause food riots and social unrest. Growing scarcities of water, land, and fuel are in all probability to position increasing pressure on food prices, even whereas not world climate change. Where these scarcities are combined by the results of world climate change, the introduction of mitigation practices that make land-use competition and so the attribution important to environmental services to mitigate world climate change, they have the potential to cause necessary changes in relative value for numerous food productions, associated an overall increase among the worth of a median food basket for the client, with concomitant can increase in value volatility. Food accessibility may be alive of pliability to secure entitlements, that is outlined as a result of the set of resources (including legal, political, economic, and social) that a private needs obtaining access to food (FAO, 2003a). Until the 70s, food security was connected in the main to national food production and world trade (Devereux and Maxwell, 2001), but since then the concept has swollen to include households' and individuals' access to food.

The mere presence of adequate display does not ensure that a private can get and consume food-that person ought to initially have access to the food through his/her entitlements. The enjoyment of entitlements that verify people's access to food depends on allocation mechanisms, affordability, and cultural and personal preferences for specific nutrient. Increased risk exposure succeeding from world climate change will reduce people's access to entitlements and undermine their food security (Killmann, 2008).

Food utilization refers to the utilization of food and also the manner a private is prepared to secure essential nutrients from the food consumed. It encompasses, the nutrition revalue of the diet, along with its composition and designs of preparation; the social values of foods, that dictate what sorts of food must be served and consumed at completely different times of the year and on different occasions; and so the standard and safety of the food provide, which could cause loss of nutrients among the food and so the unfold of food-borne diseases if not of a spare commonplace. Climate is in all probability to bring every negative and positive change in dietary patterns and new challenges for food safety, which might have a control on nutrition separate in varied ways in which. Food system stability is about by the availability of, and accessibility to, food. In long-distance food chains, storage, processing, distribution, and promoting processes contain in-built mechanisms that have protected the global food system from instability in recent times. However, if projected it can increase in weather variability happen, they are in all probability to steer to an increase among the frequency and magnitude of food emergencies that neither the global food system nor affected native food systems are adequately prepared (FAO, 2008).

6.16 POTENTIAL IMPACTS OF CLIMATE CHANGE ON FOOD AVAILABILITY

Production of food and other agricultural goods may sustain with add demand, but there are in all probability to be necessary changes in native cropping patterns and farming practices. There has been lots of analysis on the impacts that world climate change would possibly on agricultural production, notably cultivated crops. Some half of total crop production comes from forest and mountain ecosystems along with all tree crops, whereas crops cultivated on open, productive flat land account for fewer than 13% of annual world crop production. Production from every rain-fed and irrigated agriculture in dryland ecosystems accounts for regarding 25%, and rice produced in coastal ecosystems for regarding 12% (Millennium Theme Assessment, 2005).

The valuation of world climate change impacts on agricultural production, food provides and agriculture-based livelihoods ought to take into thought the characteristics of the agro-ecosystem where specific climate-induced changes in biochemical processes are occurring, so as to see the extent thereto such changes are positive, negative or neutral in their effects. The question ready greenhouse fertilization effect will manufacture native helpful effects where higher levels of atmospherically carbon dioxide gas stimulate plant growth. This could be expected to occur primarily in temperate zones, with yields expected to increase by 10 to 25% for crops with a lower rate of natural action efficiency (C3 crops), and by 0 to 10% for those with a more robust rate of natural action efficiency (C4 crops), assumptive that carbon dioxide gas levels among the atmosphere reach 550 elements per million (IPCC, 2007c); these effects are not probability to influence projections of global food display, however (Tubiello et al., 2007).

Mature forests are also not expected to be affected, although the growth of young tree stands is raised (Norby et al., 2005). The impacts of mean temperature increase are expected otherwise reckoning to the location (Leff et al., 2004). As an example, moderate warming (increases of 1°C to 3°C in mean temperature) is supporting crop and pasture yields in temperate regions, whereas in tropical and seasonally dry regions, it's in all probability to possess negative impacts, notably for cereal crops. Warming of quite 3°C is anticipated to possess negative effects on production in all regions (IPCC, 2007c).

The provision of meat and different livestock product are influenced by crop production trends, as feed crops account for roughly 25% of the world's cropland. For climate variables like rainfall downfall, soil condition, temperature, and radiation, crops have thresholds on the far side that growth and yield are compromised (Porter and Semenov, 2005). As an example, cereals, and flowering tree yields are injured after two days of high temperatures or lowest an exact threshold (Wheeler et al., 2000). Among the EU wave of 2003, once temperatures were 6°C higher than long-term suggests, crop yields decreased by 36% for maize in Italy, and by 25% for fruit and 30% for forage in France (IPCC, 2007c).

Raised intensity and frequency of storms, altered hydrological cycles, and precipitation variance even have long-term implications on the viability of current world agroecosystems and future food system accessibility. Wild foods are notably necessary to households that struggle to provide food or secure associate gain. A modification among the geographic distribution of wild foods succeeding in ever-changing rainfall downfall and temperatures changing could so have a sway on the availability of food. Changes in climate have a diode to special declines among the availability of intractable foods by a spread of ecosystems, and any impacts are expected as a result of the global climate continues to change. For the 5,000 plant species examined in an exceedingly desert African study (Levin and General, 2005), it's foretold that 81 to 97% of the suitable habitats will decrease in size or shift because of world temperature change.

By 2085, between 25 and 42% of the species' habitats are expected to be lost altogether. The implications of these changes are expected to be notably large among communities that use the plants as food or medication. Constraints on water availability are a growing concern that world temperature change will exacerbate. Conflicts over water resources will have implications for every food production and people's access to food in conflict zones (Gleick, 1993). Prolonged and continual droughts can cause loss of productive assets that undermine the property of resource systems supported rain-fed agriculture. As an example, drought, and deforestation can increase blaze danger, with resultant loss of the vegetative overlay needed for grazing and fuelwood (Laurence and Williamson, 2001). In Africa, droughts can have severe impacts on livestock.

Food production varies spatially, so food must be distributed between regions, the most agricultural production regions are characterized relatively stable climate, but many food-insecure regions have extraordinarily variable climates. The foremost grain production regions have a for the foremost half continental climate, with dry or a minimum of atmospheric condition throughout time of year, that allows the bulk handling of harvested grain whereas not special infrastructure for cover or immediate treatment looking forward to the prevailing temperature regime, however, a modification in climate through increased temperatures or unstable, damp climate could result in grain being harvested with quite the 12 to 14% condition required for stable storage. Because of the amounts of grain and general lack of drying facilities in these regions, this would possibly produce hazards for food safety, or even cause complete crop losses, following from contamination with microorganisms and their metabolic product. It might result in a rise in food costs if the stockiest have to be compelled to invest in new storage technologies to avoid the matter. Distribution depends on the responsibility all-important capability, the presence of food stocks and once necessary-access to food aid (Maxwell and Isopod, 2003).

These factors' success usually depends upon the pliability to store food. Storage is filled with ways at the national level and by physical infrastructure at the native level. Transport infrastructure shorted food distribution in many developing countries, where infrastructure affected by climate, through either heat stress on roads or increased frequency of flood events that destroy infrastructure, there are impacts on food distribution, influencing people's access to markets to sell or purchase food (Abdulai and Crole Rees, 2001).

Exchange of food takes place at all levels-individual, household, community, regional, national, and world. At the bottom levels, exchanges typically take the shape of reciprocal cordial reception, gift-giving or barter, and functions a really necessary mechanism for handling provide fluctuations. If the ever-changing climate produces trend declines in native production, the potential of affected households to possess interaction in these ancient types of exchange is maybe going to say no. Trade remains the foremost mechanism for exchange in today's world economy. Although most food trade takes, place at intervals a national border, world trade is that the leveling mechanism that keeps exchange flowing smoothly (Stevens et al., 2003).

The relatively low worth of oceans compared with land transport makes it economically advantageous for several countries to admit international food trade to disembarrass fluctuations in domestic food provides. Where trade is heavily regulated, as in the southern continent, farmers' behavior illustrates this principle. Once a food crisis like that in southern continent in 2002, while recovery programs cause a bumper harvest of maize, in some countries the maize may not notice its approach into national grain markets, as declared or anticipated producer prices and market laws could encourage farmers to channel their surplus outside formal markets (Mano et al., 2003).

FAO comes that the impacts of world climate change on world crop production are slight up to 2030. Later year, however, widespread declines among the extent and potential productivity of cropland could occur, with variety of the severest impacts in all probability to be felt among the presently food-insecure areas of SSA, that have with the littlest quantity ability to adapt to world climate change or to compensate through larger food imports (FAO, 2003b). Although the projections advised that normal carryover stocks, food aid, and international trade must be able to address the localized food shortages that are in all probability to result from crop losses due to severe droughts or floods, this could be presently being questioned seeable of the growth that the world has practiced since 2006. In line with the FAO of the UN, the global food price index rose by 9% in 2006 and by 37% in 2007. The value boom has been within the thick of lots of upper price volatility than among the past, notably among the cereals and oilseeds sectors, reflective reduced inventories, strong relationships between agricultural trade and different markets, and so the prevalence of larger market uncertainty usually. This has triggered a widespread concern regarding food price inflation, that's fuelling debates regarding the long-term direction of agricultural trade prices in importation and exportation countries, be they producer or poor, and giving rise to fears that a world food crisis similar in magnitude to those of the first 70s and 80s may even be close, with little prospect for a quicker sure as a result of the results of world climate change take their toll.

6.17 POTENTIAL IMPACTS OF CLIMATE CHANGE ON FOOD ACCESS

Allocation: Food is distributed through markets and non-market distribution mechanisms. Factors that verify whether or not people will have access to spare food through markets are thought of among the subsequent part on affordability. These factors embrace income-generating capability, amount of remuneration received for product and merchandise sold or labor and services rendered, and so the quantitative relation of the worth of a minimum daily food basket to the common daily gain. Non-market mechanisms embrace production for own consumption, food preparation, and allocation practices at intervals the home, and public or charitable food distribution schemes. For rural those that manufacture a substantial part of their own food, world climate change impacts on food production may reduce availability to the aim that allocation selections have to be compelled to be created at intervals the house. A family may reduce the daily amount of food consumed equally among all house members, or assign food preferentially to certain members, usually the healthy male adults, UN agency is assumed to would like it the foremost to stay work and continue operational to take care the family. Non-farming low-income rural and civilized households whose incomes fall below the poverty line because of world climate change impacts will face similar selections. Urbanization is increasing quickly worldwide, and a growing proportion of the increasing urban population is poor (Ruel et al., 1998). Allocation issues succeeding from world climate change are so in all probability to become lots of and lots of necessary in urban areas over time. Where urban gardens are prepared, they supply farming manufacture for home use and the native sale, though urban land-use restrictions and so the rising worth of water and land restrain their potential for growth. Urban agriculture faces a restricted ability to contribute to the welfare of poor people in developing countries as results of the bulk of their staple food require-ments still have to be compelled to be transported from rural areas (Ellis and Sumberg, 1998).

Political and social power relationships are key factors influencing allocation picks in times of lack. If agricultural production declines and households realize different support activities, social processes and reciprocal relations throughout that domestically created food is given to different members of the family in exchange for his or her support may modification or disappear altogether. Public and charitable food distribu-tion schemes apportion food to the foremost indigent, but are subject to public perceptions regarding world organization agency wishes facilitate, and social values regarding what moderately facilitate its obligatory lots of moneyed segments of society to provide. If world climate change creates alternate lots of imperative claims on public resources, support for food distribution schemes may decline, with resultant can increase among the incidence of food insecurity, hunger, and famine connected deaths In many countries, the quantitative relation of the value of a minimum daily food basket to the common daily gain is utilized as a live of financial condition (World Bank Poverty Net, 2008).

Once this quantitative relation falls below an exact threshold, it signifies that food is reasonable and people are not impoverished; once it exceeds the established threshold, food is not low cost and people are having issues obtaining enough to eat. This criterion is an associate indicator of chronic financial condition, and would possibly even on the far side to verify once people have fallen into temporary food insecurity, because of reduced food provide and increased costs, to an unforeseen fall in household gain or to every. Income-generating capability and so the payment received for product and merchandise sold or labor and services rendered are the primary determinants of average daily gain. The incomes of all farming households depend upon what they acquire from promoting some or all of their crops and animals each year. Commercial farmers are generally protected by insurance, but small-scale farmers in developing countries are not, and their incomes can decline sharply if there is a market glut, or if their own crops fail which they do not have something to sell once prices are high.

Most food is not made by individual households but obtained through buying, commerce, and borrowing (Toit and Ziervogel, 2004). Climate impacts on income-earning opportunities will have an effect on the pliability to buy for food and a modification in climate or climate extremes may have a control over the availability of certain nutrients, which might influence their price. High costs may make sure foods unaffordable and might have a sway on individuals' nutrition and health. Changes among the demand for seasonal agricultural labor, caused by changes in production practices in response to world climate change, can have a control on income-generating capability fully or negatively. Mechanization could do decrease the requirement for seasonal labor in many places, and labor demands are usually reduced once crops fail, for the most part, because of such factors as drought, flood, frost or cuss outbreaks, which could be influenced by climate. On the reverse hand, some adaptation selections increase the demand for seasonal agricultural labor. Native food prices in most zones of the world are powerfully influenced by world market conditions; yet there may even be short fluctuations connected to variation in national yields that are influenced by climate, among different factors. An increase in food costs options a true gain impact, with low-income households usually suffering most, as they need an inclination to devote larger shares of their incomes to food than higher-income households do (Thomsen and Metz, 1998). Once they cannot afford food, households regulate by intake less of their preferred foods or reducing total quantities consumed as food costs increase. Given the growing sort of folks that depend upon the marketplace for their food provide, food costs are very important to consumers' food security and must be watched.

Food usually travels long distances (Pretty et al., 2005), and this has implications for costs. Increasing fuel costs could result in costlier food and increased food insecurity. The growing marketplace for biofuels is anticipated to possess implications for food security, as results of crops grown as feedstock for liquid biofuels can replace food crops that then have to be compelled to be sourced elsewhere, at higher prices. Food preferences verify the methods of food households will plan to acquire. Changing climate may have a control on every physical and so the economic availability of certain preferred food things, that may produce it impossible to satisfy some preferences. Changes in the availability and relative costs for major food things could lead on to fogeys either dynamical their food basket, or payment a bigger share of their gain on food once costs of most well-liked food things increase.

In southern Africa, as AN example, many households eat maize as the staple crop, but once there is less precipitation, sorghum fares higher, and other people could consume lots of it. Many of us like maize to sorghum, however, so still plant maize despite poor yields, and would rather purchase maize than eat sorghum, once necessary. The extent thereto to food preferences modification in response to changes among the relative prices of grain-fed beef compared with different sources of animal protein are a really necessary determinant of food security among the medium term. Increased prices for grain-fed beef are predictable, because of the increasing competition for land for intensive provender production, the increasing insufficiency of water and rising fuel costs (FAO, 2007b). If preferences shift to different sources of animal protein, the livestock sector's demands on resources that are in all probability to be below stress as a consequence of world climate change may even be contained. If not, continued growth in demand for grain-fed beef, from wealthier segments of the world's population, could trigger blanket can increase in food prices, which could have serious adverse impacts on food security for urban and rural poor (Figure 6.3).

6.18   POTENTIAL IMPACTS OF CLIMATE CHANGE ON FOOD UTILIZATION

Food insecurity is typically associated with undernourishment, as results of the diets of the people world organization, the agency is unable to

FIGURE 6.3 Policies and priorities for achieving higher competitiveness, Egypt (Hamada 2018).

satisfy all of their food wishes generally contain a high proportion of staple foods and lack the variability needed to satisfy biological process requirements. Declines among the availability of intractable foods, and limits on small-scale farming production due to insufficiency of water or labor succeeding from world climate change could have a control on biological process standing adversely. In general, however, the foremost impact of world temperature change on nutrition is maybe going to be felt indirectly, through its effects on the gain and capability to shop for a diversity of foods. The physiological utilization of foods consumed in addition affects nutritional position standing, and this-in turn-is filled with malady (IPCC, 2007b).

World climate change will cause new patterns of pests and diseases to emerge, poignant plants, animals, and humans, and move new risks for food security, food safety, and human health. Increased incidence of water-borne diseases in flood-prone areas, changes in vectors for climate-responsive pests and diseases, and the emergence of recent diseases could have a control on every the natural phenomenon and people's physiological capability to urge necessary nutrients from the foods consumed. Vector changes are a virtual certainty for pests and diseases that flourish entirely at specific temperatures and below specific humidity and irrigation management regimes. These will expose crops, livestock, fish, and humans to new risks thereto they cannot, however, acclimatize with it. They're going to in addition place new pressures on caregivers at intervals the house, world organization agency are usually women, and might challenge health care institutions to reply to new parameters. Protozoa infection notably is anticipated to change its distribution as a result of world climate change (IPCC, 2007b).

In coastal areas, lots of oldsters may even be exposed to vector- and water-borne diseases through flooding connected to lowland rise. Health risks will even be connected to changes in diseases from either increased or shrunken precipitation, lowering people's capability to utilize food effectively and sometimes resulting in the requirement for improved biological process intake (IPCC, 2007b). Where vector changes for pests and diseases are foretold, varieties associated breeds that are proof against the in all probability new arrivals are introduced as an adjective measure. A recent upsurge among the design of recent viruses may also be climate-related, though' this link is not certain. Viruses like vertebrate contagion, Ebola, HIV/AIDS, and respiratory disease have varied implications for food security, along with risk to the livelihoods of small-scale poultry operations among the case of vertebrate contagion, and so the more biological process requirements of affected people among the case of HIV-AIDS. The social and cultural values of foods consumed additionally are stuffed with the availability and affordability of food. The social values of foods are necessary determinants of food preferences, with foods that are accorded high worth being preferred, and folks accorded low worth being avoided. In many ancient cultures, feasts involving the preparation of specific foods mark necessary seasonal occasions, rites of passage, and social occasion events. The increased price or absolute inconvenience of these foods could force cultures to abandon their ancient practices, with unpredictable secondary impacts on the cohesiveness and property of the cultures themselves. In many cultures, the reciprocal giving of gifts or sharing of food is common. It's usually thought of a social obligation to feed guests; even once they need, discarded unexpectedly. In conditions of chronic food insufficiency, households' ability to honor these obligations is shattering down, and this trend is maybe visiting be bolstered in locations where the impacts of world climate change contribute to increasing incidence of food shortages (Killmann, 2008).

Food safety may even be compromised in varied ways in which. Increasing temperature may cause food quality to deteriorate, unless there is increased investment in refrigeration instrumentality or lots of reliance on the speedy method of place attributes able foods to extend their shelf-life. Shrunken water availability has implications for food method and preparation practices, notably among the semi tropics, where a switch to dry method and preparation methods may even be required. Changes in land use, driven by changes in precipitation or increased temperatures, will alter how people spend their time. In some areas, kids might need to organize food, wherever oldsters add the sector, increasing the chance that pretty hygiene practices may not be followed (Killmann, 2008).

6.19   POTENTIAL IMPACTS OF CLIMATE CHANGE ON FOOD SYSTEM STABILITY

Many crops have annual cycles, and yields fluctuate with climate variability, notably rainfall precipitation and temperature. Maintaining the continuity to food provide production is seasonal, and it is so tough. Droughts and floods are a selected threat to food stability and can produce chronic and transitory food insecurity. Every anticipated to become lots of frequent, lots of intense and fewer sure as a consequence of world temperature change. In rural areas that depend upon rain-fed agriculture for a really necessary part of their native food provide, changes among the quantity and temporal arranging of rainfall precipitation at intervals the season and rising in weather variability are probability to irritate the dangerousness of native food systems (Killmann, 2008).

The affordability of food is set by the affiliation between household gain and so the worth of a typical food basket. World food markets may exhibit larger price volatility, jeopardizing the stability of returns to farmers and so the access to purchased food of every farming and non-farming poor people. Increasing instability of provide, because of the implications of world temperature change, will presumptively cause can increase among the frequency and magnitude of food emergencies therewith the global food system is unequipped to cope. An increase in human conflict, caused part by migration and resource competition because of changing climate would even be destabilizing for food systems in the smallest levels. World climate change may exacerbate conflict in numerous ways in which, though links between world climate change and conflict must lean with care. Increasing the incidence of drought may force people to migrate from one place to another, giving rise to conflict over access to resources in the receiving place. Resource insufficiency will even trigger conflict and will be driven by the world environmental amendment (Killmann, 2008). Grain reserves are utilized in emergency-prone areas to catch au fait crop losses and support food relief programs for displaced people and refugees. Higher temperatures and humidity associated with world climate change may need increased expenditure to preserve a hold on grain, which can limit countries' ability to take care of reserves of sufficient size to retort adequately to large-scale natural or human-incurred disasters (Figure 6.4).

FIGURE 6.4   Achieving higher competitiveness in Upper Egypt, Egypt (Hamada, 2018).

6.20 LIVELIHOOD VULNERABILITY

The livelihoods perspective is sometimes used as the simplest way of labor range of sectors and also the way they need a control on individual livelihoods. Viewing food security from a livelihoods perspective makes it getable to assess the assorted components of food security holistically at the home level. Livelihoods are made public as a result of the bundle of assorted forms of assets, skills, and activities that amendment a private or house to survive (FAO, 2003a). These assets embrace physical assets like infrastructure and house items; financial assets like stocks of money, savings, and pensions; natural assets like natural resources; social assets, that are supported the cohesiveness of people and societies; and human assets that depend upon the standing of individuals and would possibly involve education and talent. These assets revise over time and are utterly different for numerous households and communities. The amounts of these assets that a house or community possesses or can merely gain access to are key determinants of property and resilience (Figure 6.5).

FIGURE 6.5 Rural income and employment benefits and how they might be enhanced (Hamada, 2018).

Marginal groups embrace those with few resources and short access to power, which could constrain people's capability to adapt to climate changes that may have a negative impact on them. It's generally people's few productive assets that are at greatest risk from the impacts of world climate change. Physical assets are suspended or destroyed, financial losses are incurred, natural assets are degraded and social assets are undermined. The modification in seasonality attributed to world temperature change can cause certain nutrients becoming scarcer at certain times of the year. Such seasonal variations in food provide, along with vulnerabilities to flooding and fireplace can make livelihoods lots of vulnerable at certain times of the year. Although these impacts may appear indirect, they are necessary as a result of many marginal resource groups are close to the financial condition margin, and food may be a key a part of their existence (FAO, 2008). Agriculture is sometimes in the middle of the resource ways of these marginal groups; agricultural employment, whether or not farming their own land or performing on that of others, is crucial to their survival. In many areas, the challenges of rural livelihoods drive urban migration. As a result of the variability of poor and vulnerable people living in urban slums grows, the availability of non-farm employment opportunities and so the access of urban dwellers to adequate food from the market will become additional necessary drivers of food security. A recent ILO study (ILO, 2005) suggests that there will be necessary variations between middle- and low-income countries among the ways in that throughout which world climate change affects agriculture-based livelihoods.

In middle-income countries, an exploitation technique looks to be delivery regarding declines in unpaid on-farm family labor and can increase in wage employment. In low-income countries, however, wage work is declining, whereas self-cultivation and mixed agreement form increase. This implies that whereas the adverse impacts of world temperature change on agricultural production in middle-income countries are lots of in all probability to be felt as loss of employment opportunities, reduction wage earnings and loss of shopping power for agricultural wage staff, in low-income countries they are probability to be felt as declines in own production for house consumption by farmer of farming households (Killmann, 2008).

Livelihood groups that warrant special attention among the context of world climate change include:

- Low-income groups in drought- and flood-prone areas with poor food distribution infrastructure and restricted access to emergency response;
- Low- to middle-income groups in flood-prone areas which can lose homes, hold on the food, personal possessions and suggests that of getting their support, notably once the water rises terribly quickly and with force, as in sea surges or flash floods;
- Farmers whose land becomes submerged or deactivated by the sealying rise or water intrusions;
- Producers of crops which cannot be probable under changing temperature and rainfall downfall regimes;
- Producers of crops in risk from high winds;
- Poor livestock curator in drylands where changes in rainfall downfall patterns can vestige on forage accessible and quality;
- Managers of forest ecosystems that provide forest product and environmental services;
- Fishers whose infrastructure for fishing activities, like port and landing facilities, storage facilities, fish ponds and operation areas becomes submerged or disrupted by sea-level rise, flooding or extreme weather events;
- Fishing communities that bank heavily on coral reefs for food and protection from natural disasters;
- Fishers/aqua farmers, who are suffering decreasing catches from shifts in fish distribution and so the productivity of aquatic ecosystems, caused by changes in ocean currents or increasing raised discharge of freshwater into oceans.

Within these resource groups, producers at utterly different points of the food chain, like fishers versus fish cleaners, would have utterly different vulnerabilities and access to header mechanisms. Producers of assorted forms of crops, like crops accessible versus those for home consumption, may face utterly completely different risks and management selections (e.g., access to irrigation water or seeds). Gender and age variations additionally can have a control on the degree of risk faced by individuals at intervals a vulnerable cluster. Agriculture-based resource systems that are already vulnerable to world temperature change face immediate risk of increased failure, loss of livestock and fish stocks, increasing water scarcities and destruction of productive assets. These systems embrace small-scale rain-fed farming, pastoralism, interior and coastal fishing/ aquaculture communities, and forest-based systems. Rural people inhabiting coasts, floodplains, mountains, drylands and so the Arctic is most in peril. The urban poor, notably in coastal cities and plain settlements, in addition, face increasing risks. Among those in peril, pre-existing socioeconomic discriminations are in all probability to be aggravated, inflicting nutritional standing to deteriorate among ladies, young kids, and senior, unwell, and disabled individuals (FAO, 2008).

Over time, the geographic distribution of risk and vulnerability is maybe a visiting shift. Future vulnerability is maybe going have a control on not entirely farmers, fishers, herders, and forest-dependent people, but in addition, low-income city dwellers, in every developed and developing countries, whose sources of resource and access to food may even be in peril from the impact of most weather events and variable food costs, and UN agency lack adequate coverage. Some agriculture-based livelihoods may have the good thing about the results of world climate change, whereas others are undermined. The resource standing of agricultural staffs additionally can modification if centers of agricultural production shift or methods of production recede effort-full in response to world climate change. All wage earners face new health risks that may cause declines in their productivity and earning power. World climate change additionally can have control over people otherwise looking forward to such factors as landownership, quality holdings, marketable skills, gender, age, and health standing (Killmann, 2008).

Fishing is usually integral to mixed support methods, throughout that people make the most of seasonal stock availability or resort to fishing once different sorts of food production and financial gain generation disappoint. Fishing is sometimes related to the extreme financial conditions and can function as a major safety net for folk with restricted resource alternatives and extreme vulnerability to changes in their surroundings. However, the viability of fishing as a property support is vulnerable by world climate change. Fishing communities that depend upon interior geographical point resources are in all probability to be notably vulnerable to climate change; access to water resources and arrangements with different sectors for sharing and utilize will become a key to the future property. World climate change is in addition in all probability to possess substantial and intensive impacts on coastal fisheries and fishing communities. Major physical impacts of world climate change on the marine system are changes in ocean currents, a rise in average temperature, sharpening of gradient structures, and large and speedy can increase of recent discharge; these usually trigger an increase in chemical nutrients, usually compounds containing component or phosphorus, resulting in lack of substance and severe reductions in water quality and in fish and different animal populations (FAO, 2008).

Biological responses to those changes are expected to be racket-like, i.e., once a threshold is reached, things shift from one section to a special. Fishing is truly a searching activity, so its success or failure depends heavily on the vagaries of nature. World climate change is creating lots of anomalies, every failure, and bonanzas, among multiple species, additionally as forceful shifts among the areas where little, migrating fish are found. Coastal people and communities that depend upon fishing in locations where a rise in water level makes relocation inevitable would force more support, as they need to not entirely migrate, but in many instances in addition notice new, unknown ways in which of earning a living (FAO, 2007a).

All IPCC emission things assume that economies for the world as a complete will still grow, albeit at utterly different rates and customarily with necessary regional variations, counting on the position (IPCC, 2000). However, it's in addition getable that the impact of world climate change will very curtail process. If world financial markets are not able to sustain with continued high losses from extreme weather events, and large numbers of individual households in developed and rising developing countries experience uncompensated declines among the worth of their personal assets and income-generating capability, world economic recession, and a deterioration among the food security situation the smallest amount bit levels is, in addition, a prospect, swing everybody in peril.

6.21 LIVING WITH UNCERTAINTY AND MANAGING NEW RISKS

Adapting to world climate change involves managing risk by up the quality of information and its use, providing insurance against world climate change risk, adopting legendary good practices to strengthen the resilience of vulnerable resource systems, and finding new institutional and technological solutions. People in the insurance business originate a clear distinction between certain and unsure risks: a risk is definite if the possibilities of specific states occurring among the longer-term are specifically legendary, and unsure if these prospects are not specifically legendary (Kunreuther and Michel-Kerjan, 2006).

In the field of world climate change, there are still lots of uncertainty regarding the possibilities of various getable changes occurring in specific locations. This might be treated by finance in improved data to chop back the degree of native uncertainty, or by spreading the unsure risk through some style of world insurance theme. Information regarding the long-term will invariably be unsure, but this high degree of uncertainty regarding potential native impacts of world temperature change may be reduced through science. Different priorities embrace recognizing the requirement for decision-making among the face of uncertainty, bridging the gap between scientific and ancient perceptions of world climate change, and promoting the adoption of practices that are in line with the preventive approach and adjective management principles which might strengthen the resilience and property of vulnerable resource systems. Climate-related risks have a control on everybody during a manner or another, so innovative insurance schemes sort of a world insurance fund for world climate change damage, or swollen native coverage of weather-based insurance are in all probability to be needed. No risk management policy or the program will work unless those in peril feel that it addresses their wishes, so adequate provisions ought to be created to allow the foremost vulnerable to participate opt for that actions to want to strengthen their resilience. The state of the art for these approaches and so the implications of safeguarding food security among the face of world climate change are explored (Killmann, 2008).

6.22 PROMOTING INSURANCE SCHEMES FOR CLIMATE CHANGE RISK

In 2007, the world Economic Forum created public the five core areas of worldwide risk as economic, environmental, government, social, and technological. At intervals these, world climate change is seen united of the shaping challenges for the twenty-first century, as a result of it may be a world risk with impacts manner on the so much aspect the environment (World Economic Forum, 2007). The insurance business is among the economic sectors that are already experiencing adverse impacts of world climate change. Flush countries bank heavily on the personal insurance business to defend their voters against natural disasters. In line with a recent report, these countries account for 93% of the planet's insurance market (Hamilton, 2004). This market is increasing strained as it tries to reply to astronomical will increase in claims associated related to the impacts of most weather events in North America and Europe. In the US, a 2005 study on the availability and affordability of climate risk insurance found, that weather-related losses were growing 10 times faster than every premium and so the economy, and quite 10 times faster than the population; it, in addition, noted that this trend would be combined by continued settlement in unsound areas (Mills et al., 2005). Higher losses are already leading the insurance business to charge higher premiums, raise deductibles; lowermost coverage limits, and limit the classes of natural disasters or ruinous events that can be insured. It's finished that, Given the very important role that insurance plays in America and the world economy, reduced access to low-cost insurance would have profound impacts on each of customers and businesses (Mills et al., 2005).

Although this statement refers to the American economy, it's equally applicable all over among the planet, and so the implications for future food security are probably very serious. Typical types of coverage for weather- and climate-related events (e.g., floods, windstorms, thunderstorms, hailstorms, ice storms, wildfires, droughts, heat waves, lightning strikes, subsidence damage, and coastal erosion) embrace coverage for property damage, business interruptions, and loss of life or limb. If climate stresses cause the insurance business among the developed world to forestall providing such coverage once natural disasters are involved, many previously food-secure people are exposed to special uncompensated losses of property and suggests that of resource, which will plunge them into a state of vulnerability that has previously been associated in the main with developing countries. Increasing climate stresses and so the retreat of the private sector insurance business from covering losses caused by ruinous natural events will cause increasing desires national and native governments to step in. Most governments already operate public sector insurance programs for major risks if there is no personal sector coverage, like for crop loss, flood, and earthquakes damage; they are help them in disaster state and recovery operations. These programs are also experiencing increasing losses, however, thus the financial burden of maintaining this social safety net protection among the face of more demands generated by the impacts of world climate change may even be on the so much aspect what many governments in developed countries can afford. As a result of there is little personal sector insurance in developing countries, different approaches to insurance have evolved to accommodate low-income groups. Informal, domestically primarily based small insurance initiatives provide a most well-liked different as a result of the premiums are low and so the foundations are usually less stringent than for industrial insurance (Hashemi and Foose, 2007). Public personal partnerships are also additional and additional widespread, and sometimes involve the govt. Coordinating and/or adding to premium payments created by those to be insured. As climate-related risks have a control on everybody, insurance against the implications of ruinous weather events must be globalized, and costs diminished through action to mitigate world climate change.

6.23   CONCLUSION

This chapter provides a brief overview of background of climate change and food security wherein global food production must growth with the aid of 50% to fulfill the projected demand of the world's populace by way of 2050. Meeting this hard project might be made even tougher if climate change melts portions of the Himalayan glaciers to affect 25% of world cereal production in Asia by influencing water availability. Pest and disease control has played its function in doubling food production inside the final 40 years, but pathogens still claim 10–16% of worldwide harvest. We recollect the effect of climate change on the many complex organic interactions affecting pests and pathogen influences and how they are probably manipulated to mitigate these effects. Integrated answers and worldwide co-ordination of their implementation are taken into consideration essential. Either, too this chapter provides a brief overview of background of climate change and food security, food security, food system, food chain, climate and its measurement, climate system, climate variability and climate change, effects of global warming on the climate system, acclimatization, adaptation and mitigation, agriculture, climate and food security, food security and climate change: a conceptual framework, vulnerability to climate change, food system vulnerability, potential impacts of climate change on food availability, potential impacts of climate change on food access, potential impacts of climate change on food utilization, potential impacts of climate change on food system stability, livelihood vulnerability, Living with uncertainty and managing new risks and promoting insurance schemes for climate change risk.

KEYWORDS

- **climate change and food security (CCFS)**
- **domestic nourishment systems**
- **drylands**
- **food security**
- **precipitation**
- **vulnerability**

REFERENCES

Abdulai, A., & Crole, R. A., (2001). Constraints to income diversification strategies: Evidence from Southern Mali. *Food Policy, 26*(4), 437–452.

Alexandrov, V. A., & Hoogenboom, G., (2000). Vulnerability and adaptation assessments of agricultural crops under climate change in the southeastern, USA. *Theor. Appl. Climatol., 67,* 45–63.

Cerri, C. E. P., Sparovek, G., Bernoux, M., Easterling, W. E., Melillo, J. M., & Cerri, C. C., (2007). Tropical agriculture and global warming: Impacts and mitigation options. *Sci. Agric. (Piracicaba, Braz.), 64,* 83–89.

Chen, D. X., Hunt, H. W., & Morgan, J. A., (1996). Responses of a C3 and C4 perennial grass to CO_2 enrichment and climate change: Comparison between model predictions and experimental data. *Ecological Modeling, 87,* 11–27.

Du Toit, A., & Ziervogel, G., (2004). Vulnerability and food insecurity: Background concepts for informing the development of a national FIVIMS for South Africa. Available at: www.agis.agric.za/agisweb/FIVIMS_ZA (accessed on 4 January 2020).

Ellis, F., & Sunberg, J., (1998). Food production, urban areas, and policy responses. *World Development, 26*(2), 213–225.

FAO, (1996). *Rome Declaration and the World Food Summit, Plan of Action.* Rome. Available at: www.fao.org/docrep/003/X8346E/x8346e02.htm#P1_10 (accessed on 4 January 2020).

FAO, (2003a). In: Stamoulis, K., & Zezza, A., (eds.), *Conceptual Framework for National, Agricultural, Rural Development, and Food Security Strategies and Policies.* Rome.

FAO, (2003b). *World Agriculture: Toward 2015/2030.* Chapter 13, Rome, Earth scan.

FAO, (2007a). *Building Adaptive Capacity to Climate Change: Policies to Sustain Livelihoods and Fisheries.* New directions in fisheries: A series of policy briefs on development issues no. 08. Rome.

FAO, (2007b). *Food Outlook – November 2007: High Prices and Volatility in Agricultural Commodities.*

FAO, (2007c). *National Programs for Food Security: FAO's Vision of a World Without Hunger.* Rome.

FAO, (2008). Food and agriculture organization of the United Nations, Rome. *Climate Change and Food Security: A Framework Document.*

Gleick, P. H., (1993). *Water in Crisis: A Guide to the World's Fresh Water Resources.* New York, Oxford University Press.

Grasty, S., (1999). Agriculture and climate change: In Auger, P., & Suwanraks, R., (eds.), *TDRI Quarterly Review, 14*(2), 12–16.

Gunderson, C. A., Edwards, N. T., Walker, A. V., O'Hara, K. H., Campion, C. M., & Hanson, P. J., (2012). Forest phenology and a warmer climate – growing season extension in relation to climate provenance: In: *Global Change Biology, 18*, 2008–2025.

Hamada, Y. M. (2018). Special pictures album, Took by Youssef M. Hamada, Egypt, 2018.

Hamilton, K., (2004). Insurance and financial sector support for adaptation. In: *Climate Change and Development, IDS Bulletin* (Vol. 35, No. 3). Brighton, UK, Institute of Development Studies.

Hashemi, S., & Foose, L., (2007). *Beyond Good Intentions: Measuring the Social Performance of Microfinance Institutions.* Focus note no. 41. Consultative GROUP to assist the poor (CGAP), Available at: www.microfinancegateway.org/files/42300_file_ FocusNote_41.pdf (accessed on 16 January 2020).

ILO, (2005). In: Majid, N., (ed.), *On the Evolution of Employment Structure in Developing Countries.* Employment strategy papers No. 18. Available at: www.ilo.org/public/ english/employment/strat/download/esp2005–18.pdf (accessed on 4 January 2020).

ILO, (2007). Chapter 4: Employment by sector. In: *Key Indicators of the Labor Market, KILM* (5th edn.). Available at: www.ilo.org/public/english/employment/strat/kilm/ download/kilm04.pdf (accessed on 4 January 2020).

IPCC, (2000). *Special Report on Emissions Scenarios.* Cambridge, UK, Cambridge University Press.

IPCC, (2007a). Summary for policymakers. In: Metz, B., Davidson, O. R., Bosch, P. R., Dave, R., & Meyer, L. A., (eds.), *Climate Change 2007: Mitigation: Contribution of Working Group III to the Fourth Assessment Report of the Intergovernmental Panel on Climate Change.* Cambridge University Press, Cambridge, United Kingdom and New York, NY, USA.

IPCC, (2007b). *Climate Change 2007: Impacts, Adaptation, and Vulnerability.* Contribution of Working Group II to the Fourth Assessment Report of IPCC, Cambridge, UK, Cambridge University Press.

IPCC, (2007c). *Climate Change 2007: The Physical Science Basis.* Contribution of Working Group I to the Fourth Assessment Report of IPCC, Cambridge, UK, Cambridge University Press.

Kunreuther, H. C., & Michel-Kerjan, E. O., (2006). *Climate Change, Insurability of Large-Scale Disasters and the Emerging Liability Challenge.* Paper prepared for the University of Pennsylvania Law Conference on climate change. Philadelphia, PA, USA. Available at: www.climateandinsurance.org/news/ClimateChangePaperforPennLawReview.pdf (accessed on 16 January 2020).

Laurence, W. F., & Williamson, G. B., (2001). Positive feedbacks among forest fragmentation, drought and climate change in the Amazon. *Conservation Biology, 28*(6), 1529–1535.

Leff, B., Ramankutty, N., & Foley, J., (2004). Geographic distribution of major crops. Across the world, Article No. GB1009. *Global Biogeochemical Cycles, 18*(1).

Levin, K., & Pershing J., (2005). *Climate Science 2005: Major New Discoveries.* WRI, Issue Brief, Washington, DC.

Lukac, M., Calfapietra, C., Lagomarsino, A., & Loreto, F., (2010). Global climate change and tree nutrition: Effects of elevated CO_2 and temperature: Tree physiology [0829–318X]. *Lukac yr., 30*, 1209–1220.

Luxmoore, R. J., Wullschleger, S. D., & Hanson, P. J., (1993). Forest responses to CO_2 enrichment and climate warming. *Water, Air, and Soil Pollution, 70*, 309–323.

Mano, R., Isaacson, B., & Dardel, P., (2003). *Policy Determinants of Food Security Response and Recovery in the SADC Region: The Case of the 2002 Food Emergency.* Paper presented to the regional dialogue on agricultural recovery, Food Security and Trade Policies in Southern Africa. Gaborone, Botswana.

Maxwell, S., & Slater R., (2003). Food policy old and new. *Development Policy Review, 21*(5/6), 531–553.

Millennium Ecosystem Assessment, (2005). *Ecosystems and Human Well-Being: Synthesis.* Washington DC, Island Press for WRI.

Mills, E., Roth, R. J., & Lecomte, E., (2005). *Availability and Affordability of Insurance Under Climate Risk: A Growing Challenge for the U.S.A White Paper Commissioned by Ceres* (pp. 2, 3). Available at: www.ceres.org/pub/docs/Ceres_insure_climatechange_120105. pdf (accessed on 16 January 2020).

Morison, J. I. L., & Lawlor, D. W., (1999). Interaction between increase CO_2 concentration and temperature on plant growth. *Plant Cell and Environment, 22*, 659–682.

Norby, R. J., De Lucia, E. H., Gielen, B., Calfapietra, C., Giardina, C. P., King, J. S., et al., (2005). Forest response to elevated CO_2 is conserved across a broad range of productivity. *Proceedings of the National Academy of Sciences, 102*(50), 18052–18056.

Pleijel, H., & Uddling, J., (2011). Yields vs. quality trade-offs for wheat in response to carbon dioxide and ozone. *Global Change Biology, 18*, 596–605.

Porter, J. R., & Semenov, M. A., (2005). Crop responses to climatic variation. *Philosophical Transactions of the Royal Society B: Biological Sciences, 360*, 2021–2035.

Pretty, J. N., Ball, A. S., Lang, T., & Morrison, J. I. L., (2005). Farm costs and food miles: An assessment of the full cost of the UK weekly food basket. *Food Policy, 30*(1), 1–19.

Qaderi, M. M., & Reid, D. M., (2009). Crop responses to elevated carbon dioxide and temperature (Chapter 1). In: Singh S. N., (ed.), *Climate Change and Crops, Environmental Science and Engineering.* Springer-Verlag Berlin Heidelberg. doi: 10.1007/978–3-540–88246–6-1.

Roos, J., Hopkins, R., Kvarnheden, A., & Dixelius, C., (2010). The impact of global warming on plant diseases and vectors in Sweden: *Eur. J. Plant Pathol., 129*, 9–19.

Ruel, M. T., Garrett, J., Morris, S., Maxwell, D., Oshaug, A., Engele, P., Menon, P., Slack, A., & Haddad, L., (1998). *Urban Challenges to Food and Nutrition Security: A Review of Food Security, Health and Care Giving in the Cities.* IFPRI FCND Discussion Papers No. 51. Washington, DC.

Stevens, C., Devereux, S., & Kennan, J., (2003). *International Trade, Livelihoods, and Food Security in Developing Countries*. IDS Working Paper No. 215. Brighton, UK, Institute of Development Studies.

Thomsen, A., & Metz, M., (1998). *Implications of Economic Policy for Food Security: A Training Manual*. Rome, FAO and the German Agency for Technical Cooperation (GTZ).

Thuiller, W., Albert, C., Araujo, M. B., Berry, P. M., Cabeza, M., Guisan, A., et al., (2008). Predicting global change impact on plant species distributions: Future challenges. In: *Perspectives in Plant Ecology, Evolution and Systematic* (Vol. 9, pp. 137–152). Rubel Foundation, Published by Elsevier GmbH.

Tubiello, F. N., Amthor, J. A., Boote, K., Donatelli, M., Easterling, W. E., Fisher, G., Gifford, R., Howden, M., Reilly, J., & Rosenzweig, C., (2007). Crop response to elevated CO_2 and world food supply. *European Journal of Agronomy, 26*, 215–228.

UNFCCC, (1992). *UN Framework Convention on Climate Change*. Article 1: Definitions. Available at: www.unfccc.int/resource/docs/convkp/conveng.pdf (accessed on 4 January 2020).

Uprety, D. C., Sen, S., & Dwivedi, N., (2010). Rising atmospheric carbon dioxide on grain quality in crop plants. *Physiology Mol. Biol. Plants*.

Wheeler, T. R., Crawford, P. Q., Ellis, R. H., Porter, J. R., & Vera, P. P. V., (2000). Temperature variability and the yield of annual crops. *Agriculture, Ecosystems and Environment, 82*, 159–167.

WMO, (1992). *International Meteorological Vocabulary* (2nd edn.). Publication No. 182, Available at: http://www.meteoterm.wmo.int/meteoterm/ns?a=T_P1.start&u=&direct=y es&relog=yes#expanded (accessed on 4 January 2020).

World Bank Poverty Net, (2008). *Measuring Poverty*. Available at: http://go.worldbank.org/VCBLGGE250 (accessed on 4 January 2020).

World Economic Forum, (2007). *Global Risks 2007: A Global Risk Network Report*, Geneva.

Wulf, K., (2008). *Climate Change and Food Security: A Framework Document*. Food and agriculture organization of the United Nations, Rome.

CHAPTER 7

Future Threats to Agricultural Food Production

ABSTRACT

Without describing the whole image of the challenge, we have a tendency to tend to cannot expect our response to match its scale, which, we are able to not wait from speaking out on the silent suffering of millions worldwide. Polls previously show that people generally are worried regarding action. Societies on the climate frontlines presently see and feel the modification. But awareness regarding the results of action is little, mostly between the poor. In industrial countries, climate change continues to be thought of a completely ecological problem. It is completed as a far danger which can disturb our future. A viewpoint bolstered by photos of glaciers and glacial bears—not human. And as yet Australia is witnessing a full decade of drought, huge areas of the US are unprotected from stronger tempests and stark water lacks—leading to crop hurt, job injuries, fires, and decease, we have to ought here to the face of these risky drawbacks. The first knockout and worst full of temperature change are that the world's humblest groups, 99% of all victims happen in developing countries. A stark variation to the straightforward fraction of world emissions due to some 50 of the smallest industrial states, if all countries were polluted slight, there would be no temperature change (Annan, 2009).

The effects of greenhouse gasses driven by economic development in some areas of the world are presently driving numerous people into poorness elsewhere. At the identical time, decades-old aid pledges still go unmet. The Millennium Growth Goals are scarce and thus the poor lack capacities to make their voices detected in worldwide arenas, or attractiveness public and private investment. For those living on the brink of survival, climate change is also a really real and dangerous hazard. For many, it is the last of deprivation. Where can a fisher go once hotter ocean temperatures eat up coral reefs Associate in nursing fish stocks? But can a weak farmer keep associate degreeimals or sow crops once the water dries up? Or families are provided for once-fertile soils and fresh are contaminated with salt from rising seas? Climate change is a broad threat, directly touching the setting, the economy, health, and safety. Many communities face multiple stresses with serious social, political, and security implications, each domestically and abroad. Uncountable folks are uprooted or for good on the move as a result. A lot of millions will follow. New climate policy ought to allow vulnerable communities to deal with these challenges. It ought to sustenance the intensive drive for a distinguished life for all, consonant with the atmosphere moreover in safety from it. Today, numerous people are presently suffering to climate change. The deadly stillness of this disaster is also a significant impediment for international action to complete it (Annan, 2009). This chapter reviews the matter of future threats to agricultural production accordingly the views of the majority of specialists in future threats to agricultural food production. Figure 7.1 shows Africa's normalized difference vegetation related to climate change.

7.1   BACKGROUND

The irrigated area of the world enhanced dramatically throughout the first and middle of the twentieth century, driven by quick growth and thus the following demand for food. Irrigation provides around 40% of the world's food, along with most of its horticultural output, from a predestined 20% of agricultural land, or regarding 300 million hectares worldwide. The Green revolution technology of high inputs of a nitrogenous fertilizer, applied to responsive short-strewed, short-season kinds of rice and wheat, usually required irrigation to grasp its potential in Asia. Public funding for irrigation development peaked at intervals the 70s, reducing to a trickle by the 90s within the aftermath of disappointment with the performance of formal huge canal systems, corruption, and rent-seeking associated with construction, and rising awareness of the impacts of large-scale water diversion on aquatic and riparian ecosystems (Burke and Faurès, 2011).

In crude terms, the Green Revolution is attributable with providing the springboard for many Asian countries to transform from farming to industrializing economies, through increasing rural wealth and aspiration. Sadly, it created a very little impact on Africa, either in terms of food security or

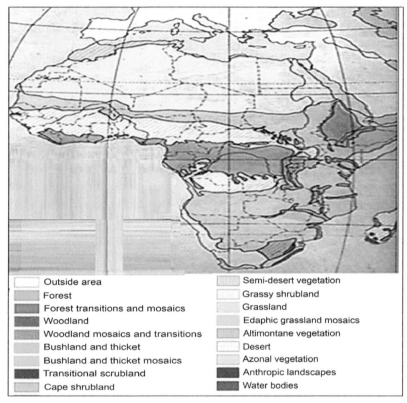

Outside area		Semi-desert vegetation	
Forest		Grassy shrubland	
Forest transitions and mosaics		Grassland	
Woodland		Edaphic grassland mosaics	
Woodland mosaics and transitions		Altimontane vegetation	
Bushland and thicket		Desert	
Bushland and thicket mosaics		Azonal vegetation	
Transitional scrubland		Anthropic landscapes	
Cape shrubland		Water bodies	

FIGURE 7.1 Africa's normalized difference vegetation related to climate change (Hamada, 2018).

wealth creation, as rural economies didn't deepen to make rural investment 'stick.' The relatively very little potential for irrigation in Africa as a whole has contributed to the current stasis. For quite 30 years the value of all major commodities reduced annually in real terms, more change the inducement to speculate public and aid finances in irrigated agriculture. However, over identical quantity personal investment in groundwater was excited by the supply of low-value pumps, power, and well construction methods, initiating at intervals the 80s and continuing rapidly in India, China, and much of geographic region. Not entirely did irrigated areas still grow, but canal irrigation had become the minor player in India by the year 2000 as individual access to groundwater services excessive. Consequently, aquifers are depleted in many parts of the world where they are most important—China, India, and thus the United States—typically fast by perverse incentives of sponsored energy and support costs for the irrigated product (Burke and Faurès, 2011).

As the world population heads for quite 9 billion folks by 2050 (under medium growth projections), the world is quickly turning into urban and wealthier. Food preferences are driving to mirror this, with declining trends at the consumption of staple carbohydrates, and an increase in demand for luxury merchandise—milk, meat, fruits, and vegetables that are heavily dependent on irrigation in many parts of the world. The efficiency of animal product is under for crops then more primary production from pastures vary lands and productive farming is needed to satisfy food demands. Future international food demand is anticipated to increase by some 70% by 2050, but will around double for developing countries. All various things being equal (that could be a world while not climate change), the quantity of water withdrawn by irrigated agriculture will increase by 11% to match the demand for biomass production (Burke and Faurès, 2011).

The long downward trend in trade goods prices created associate degree abrupt turnaround in 2007–2008 once a combination of run-down strategic reserves, poor harvests, droughts, and a sudden rush to plant biofuels in the US and Europe reduced trade volumes. Prices for rice doubled and although merchandise prices have fallen back since, the fundamentals (oil worth, biofuel development associate degreed continuing rising food demand) are presently expected to drive a quantity of high volatility in food prices. At intervals, the wake of this market turmoil, food security, and agricultural livelihoods have regained importance in development coming up with, although some countries like China seem ever lots of probably to balance extra agricultural development and investment with imports. The world encompasses an enormous stock of under-performing canal irrigation infrastructure, and a spirited groundwater sector that is competitively depleting its own lifeblood. Each produces a very important environmental externality, that needs to be managed. Not simply that, there reincorporate water to be reserved to occupy the care of environmental flows in quickly developing stream basins and specific to ecosystems in over-allocated ones (Burke and Faurès, 2011).

7.2   SCOPE OF THE CHAPTER

Clearly, not all crop production is dedicated to food security. Industrial crops (fiber and biofuels) and nutrient crops build no direct contribution to Calorie consumption by human beings although some industrial crop residues are used as cattle fodder. But overall, water management in crop production tends to be targeting food crops whosesoever the temporal property and irresponsibility of providing is crucial. Water management (irrigation, drain, and water conservation and control) achieves stability of crop production by maintaining soil conditions close to optimum for crop growth. Irrigation permits the cultivation of crops once the rain is erratic or lean, insures high-value, insecure agriculture from failure and has compete for a significant role in achieving national and regional food security, to boot as up individual livelihoods (Hussain, 2005). The extent and house of irrigation has grownup massively at intervals the 20th century but has depleted surface and groundwater flows, usually with severe consequences for aquatic eco-systems and people passionate about them (Emerton and Bos, 2004; FAO, 2004; Burke and Moench, 2000). It's more and more recognized; although seldom common observe that larger net socio-economic profit is obtained from maintaining the integrity of managed ecosystems (Cai et al., 2001; FAO, 2004).

In the future, food security strategies will be a lot of complicated. Higher temperatures will increase water demand, and wherever rain declines, several can get a lot of irrigation to confirm food security and maintain livelihoods. At identical time, water provides obtainable for irrigation can become a lot of variables and can decline in many parts of the globe. New agricultural demands are additional tempered by the requirement to attain higher equity in access to reliable food provides than within the past. As irrigation has been practiced on solely 20% of the world's tilth, there are several, typically the poorest, who have disregarded on its advantages. The requirement to take care of viable aquatic eco-systems can place additional stress on water resources, particularly wherever the poorest are obsessed with them for his or her livelihoods. Water allocations to agriculture could fall in several parts of the globe as a result of the combined impacts of climate change, environmental wants, and competition from higher worth economic sectors. There'll be sturdy pressure to provide a lot of with less water, and to unfold the advantages of all water use a lot of wide and sagely. This task is even tougher as a result of higher temperatures can cut back potential land and water productivity. These are not academic considerations. Climatic variability in south-eastern Australia has had a lot of profound impacts on water allocations and associated livelihoods in agriculture than even the foremost prudent farmers had anticipated and large changes lie ahead. However, this is an economy with alternatives: if this magnitude of amendment happens in developing countries, the impacts on poorness are expected to be rather more profound (Sperling, 2003). Temperature change can alter the productivity of aquatic ecosystems and therefore the services they supply in vital ways that, each directly, for instance in changes in rain patterns and rising ocean levels, and indirectly, through shifts in demand and trade of commodities.

7.3   CLIMATE CHANGE THROUGH THE HUMAN LENS

Climate change already encompasses a severe human impact these days, but it is a silent crisis—it is a neglected areas of analysis as a result of the temperature change dialogue has been heavily targeted on physical effects at intervals the semi-permanent. This human impact report: action, therefore, breaks new ground. It focuses on human impact rather than physical consequences. It's at the lot of and a lot of negative consequences that people around the world face as results of a driving climate. Rather than specializing in environmental events in 50–100 years, the report takes a singular social angle. It seeks to spotlight the magnitude of the crisis at hand at intervals the hope to steer the arguing towards imperative action to beat this challenge and reduce the suffering it causes. The human impact of action goes on immediately—it desires imperative attention. Events like weather-related disasters, geological process, and rising ocean levels, exacerbated by action, have a sway on folks and communities around the world. They convey hunger, disease, poverty, and lost livelihoods—reducing process and movement a threat to social and even political stability. Several folks do not appear to be resilient to extreme weather patterns and climate variability. They are unable to protect their families, livelihoods, and food provides from negative impacts of seasonal rain leading to floods or water deficiency throughout extended droughts. Action is multiplying these risks. Today, we have a tendency to tend to are at an important juncture—merely when the Copenhagen summit where negotiations for a post-2012 climate agreement ought to be finalized. Negotiators cannot afford to ignore this impact of action on human society. The responsibility of states in Copenhagen is not entirely to contain a major future threat, but to boot to cope with a significant up to now crisis. The urgency is all the lots of apparent since consultants are constantly correcting their own predictions regarding action, with the result that action is presently thought of to be occurring earlier than even the foremost aggressive models recently prompt. The unsettling anatomy of the human impact of action cannot be neglected at the negotiating tables (Fust, 2009).

7.4 CLIMATE CHANGE IS A MULTIPLIER OF HUMAN IMPACTS AND RISKS

Climate change is already seriously touching many dissimilar people nowadays and in the subsequent 20 years, those affected will probably quite double—making it the most effective rising humanitarian challenge of our time. Those seriously affected are in wish of immediate facilitate either following a weather-related disaster, or as a result of livelihoods are severely compromised by temperature change. The magnitude of those severely full of action is sort of 10 times larger than as an example of those contusions in traffic accidents each year, and quite the worldwide annual vary of recent new malaria cases. Among consecutive 20 years, 1 in 10 of the world's gift population is also directly and seriously affected. Already nowadays, several thousands of lives are lost annually due to temperature change. This might rise to roughly half a million in 20 years. Over 9 in 10 deaths are related to gradual environmental degradation due to temperature change—principally malnutrition, diarrhea, malaria, with the remaining deaths being joined to weather-related disasters drove to by temperature change (Fust, 2009).

Economic losses due to climate change presently amount to quite 100 billion North American country bucks annually that's quite the individual national GDPs of three quarters of the world's countries. This figure constitutes quite the complete of all Official Development facilitate in an extremely given year. Already nowadays, over 0.5 a billion people are at extreme risk to the impacts of climate change, and 6 in 10 people are liable climate change in an extremely physical and socio-economic sense. The majority of the world's population does not have the aptitude to deal with the impact of climate change whereas not suffering a probably irreversible loss of prosperity or risk of loss of life. The populations most gravely and at once in danger sleep in a very range of the poorest areas that are also very in danger of temperature change—in particular, the semi-arid ground belt countries from the Sahara Desert to the center East, to boot as sub-Saharan Africa (SSA), waterways, and little Island Developing States (Fust, 2009).

7.5 A QUESTION OF JUSTICE

It is a grave international justice concern that people who suffer most from climate change have done the smallest amount to cause it. Developing countries bear over 9-tenths of the climate change burden: 98% of the seriously affected and 99% of all deaths from weather-related disasters, aboard over 90% of the complete economic losses. The 50 least developed countries (LDCs) contribute, however, a simple fraction of world carbon emissions. Climate change exacerbates existing inequalities faced by vulnerable groups notably women, children, and thus the aged. Climate change exacerbates existing inequalities moon-faced by vulnerable teams significantly ladies, youngsters, and therefore the aged. Individual and social factors make sure vulnerability and capability to adapt to the results of climate change. Ladies account for a simple fraction of the world's poor and comprise 7 in 10 agricultural employees. Women and kids are disproportionately delineated among people displaced by extreme weather events and completely different climate shocks. The poorest are the hardest hit, but the human impact of action is also a world issue. Developed nations are also seriously affected, and a lot of and a lot of so (Figure 7.2). The human impact of recent heat waves, floods, storms, and forest fires in moneyed countries has been horrific. Australia is probably the developed nation most prone to the direct impacts of amendment international climate change and to boot to the indirect impact from neighboring countries that are stressed by climate changes (Wahlstrm, 2009).

7.6 THE TIME TO ACT IS NOW

Climate change threatens sustainable development and all 8-millennium development goals (MDGs). The international community united at the beginning of the new millennium to eradicate extreme hunger and poverty by 2015. Yet, today, climate change is already chargeable for forcing some 50 million more people to hungry and driving over 10 million more people into extreme poorness. Between one-fifth and one-third of Official Development, assistance is in climate-sensitive sectors and thereby very exposed to climate risks. To avert the worst outcomes of climate change, adaptation efforts should be scaled up by a factor of over 100 in developing countries. The sole path to reduce this human impact is through adaptation, but funding for adaptation in developing countries is not even a fraction of what is needed. The many-sided funds that are pledged for climate change adaptation funding presently amount to at a below place 0.5 a billion US dollars. Despite the dearth of funding, some cases of victorious adaptation do provide a glimmer of hope. The people's republic of Bangladesh is one such example. Cyclone Sidr's smitten People of Bangladesh in 2007 demonstrates but well adaptation and bar efforts pay off. Disaster preparation measures, like early warning systems and storm-proof homes, reduced injury and destruction. Cyclone Sidr's still tidy toll of 3,400, and economic damages of $1.6 billion, still compare favorably to the similar scale cyclone Nargis that hit Burma in 2008, resulting in close to 150,000 deaths and economic losses of around $4 billion (Wahlstrm, 2009).

FIGURE 7.2   Those who suffer most from climate change have done the least to cause it (Hamada, 2018).

Solutions do to boot exist for reducing gas emissions, some even with multiple benefits. For example, black carbon from soot, produce by staple energy sources in poor communities, is maybe going inflicting the most quantity as 18% of warming. The availability of cheap completely different preparation stoves to the poor can, therefore, have every positive health results, since smoke is eliminated, and an immediate impact on reducing emissions, since soot entirely remains at intervals the atmosphere for a pair of weeks. Act strategies between adaptation, mitigation, development, and disaster risk reduction can and will be reciprocally reinforcing. Action adaptation, mitigation, humanitarian facilitate and development aid underpin each other, but are supported by fully completely different sets of institutions, information centers, policy frameworks, and funding mechanisms. These policies are essential to combat the human impact of climate change, but their links to one another have received inadequate attention. A key conclusion is that the worldwide society ought to work on if humanity is to beat this shared challenge: nations should notice their common interest at Copenhagen, acting decisively with one voice; humanitarian and development actors of all sorts should pool resources, expertise, and efforts therefore on shock the quickly increasing challenges diode to by climate change; and customarily, people, businesses, and communities all over should become engaged and promote steps to tackle climate change and end the suffering it causes (Wahlstrm, 2009).

7.7 EVIDENCE THAT CLIMATE CHANGE IS A PRESENT REALITY IS UNEQUIVOCAL

Global warming is going on and human-driven emissions of greenhouse emission and completely different greenhouse gases, to boot as land-use modification, are primarily accountable. Given current trends, temperature extremes, heat waves, and important rains are expected to still intensify in every frequency and intensity, and thus the earth's temperature and seas will still rise. These conclusions lie in the middle of a 2007 report issued by the Intergovernmental Panel on climate change (IPCC), the world's foremost scientific body for the study of climate change. The IPCC was established to produce an authoritative international statement of scientific understanding of climate change. Its reports are written by a team of authors appointive by UNEP and WMO member states or authorized organizations and are supported by accord and input from many international specialists (UNEP, 2009). There is a very human tendency to need away such dire prognostications and even to question the underlying science. However, the science is presently quite firm. Folks should be told but it will have a sway on them in their country and why they need to fret regarding it presently rather than at some later time (Wahlstrm, 2009).

7.8 THE HUMAN IMPACT IS DIFFICULT TO ASSESS RELIABLY BECAUSE IT RESULTS FROM A COMPLEX INTERPLAY OF FACTORS

Climate change affects human health, livelihoods, safety, and society. To assess the human impact of climate change, report of IPCC looks the people hit by weather-related disasters like floods, droughts, and heat waves to boot as those seriously full of gradual environmental degradation like geological process and water level rise. Report of IPCC covers every the human impact of worldwide world climate change nowadays and over consecutive 20 years as this clearly demonstrates the acceleration of human impacts of climate change at intervals the near-term. The human impact continues to be powerful to assess with nice accuracy as a result of its results from an elaborate interaction of things. It's tough to isolate the human impact of action definitively from various factors like natural variability, growth, land use, and governance. In several areas, the underside of scientific proof continues to be not enough to make definitive estimates with nice truth on the human impacts of climate change. However, information and models do exist, that blood group a durable place to start for making estimates and projections which can inform disputation, political, and future analysis. This report, supported the foremost reliable knowledge base, presents estimates of the amount of people seriously affected, lives lost and economic losses due to climate change. These numbers provide the clearest gettable indication of the order of magnitude of the human impact of climate change nowadays and at intervals in the near future. A significant and aware effort has been created the human impact of action among the constraints given. Recognizing that the vital numbers are to boot significantly lower or over prompt by these estimates, they should be treated as indicative rather than definitive (Wahlstrm, 2009). Climate change goes on earlier than anyone thought accessible. Should humankind stop worrying regarding heating and instead begin panicking? The conclusion is that we have a tendency to tend to are still left with a decent probability to hold the 2°C line, still the race between climate dynamics associate degreed climate policy are a full one.

7.9 CLIMATE IMPACT WITNESS

In its Fourth Assessment Report, the IPCC found that weather patterns became extra extreme, with extra frequent and extra intense rain events and extra intense heat waves and prolonged droughts. The rhythm of weather became extra unpredictable with changes at the temporal arranging and web site of rain (Allan and Soden, 2008; Kundzewicz et al., 2007). To boot to the exaggerated severity of weather events, the sheer vary of weather-related disasters (storms, hurricanes, floods, heat waves, droughts) has quite doubled over the last 20 years (ISDRS, 2008; MRS, 2009) nowadays, the earth experiences over four hundred weather-related disasters p.a. They leave a frightening toll in their wake: nearly 90 million people requiring immediate facilitate because of personal injury, property loss, and exposure to epidemics, wellbeing or shortages of food and water. In step with believability, 219 million on the common required. Facilitate between 2000 and 2008 and fortieth is attributed to international climate change supported assumptions. The most gradual changes are rising earth surface temperatures, rising ocean levels, geological process, and changes in native rain and stream run-off patterns with exaggerated precipitation in high latitudes and shriveled precipitation in sub-tropical latitudes, salinization of stream deltas, accelerated species extinction rates, loss of diverseness and a weakening of ecosystems.

The impact of this gradual modification is tidy. It reduces access to fresh and safe drinkable, negatively affects health, and poses a real threat to food security in many countries in Africa. In some areas where employment and crop alternatives are restricted, decreasing crop yields have juncture to famines. The geologic process and completely different forms of land degradation have juncture to migration. What's a lot of, the rise in ocean levels has already spurred the first permanent displacement of small island inhabitants at the Pacific, i.e., gradual environmental degradation because of international temperature change has to boot affected semi-permanent water quality and amount in some parts of the world, and triggered can increase in hunger, insect-borne diseases like protozoa infection, completely different health problems like diarrhea and sick-nesses. It's a contributive element to misery condition, and forces folks to leave from their homes, sometimes permanently. Intuitively, if someone is suffering from water inadequacy, impoverishment, or displacement, this to boot interprets into health outcomes and food insecurity. Typically, inter-national climate change nowadays mostly affects areas already seriously suffering at a lower place the upper than mentioned factors. Likewise, health outcomes and food insecurity cause displacement and impover-ishment could that could finish in competition for scarce resources and strains on mostly already restricted government capability to require care of deteriorating conditions and would possibly ultimately cause conflict. Therefore, health outcomes and food security are taken as a result of the idea for all international climate change connected impacts, By exploita-tion this approach, the update of WHO international Burden of wellbeing study shows that future consequences of worldwide climate change have a sway on over 235 million folks nowadays (McMichael et al., 2004; WHO, 2004).

7.10 THOSE MOST VULNERABLE LIVE IN THE SEMI-ARID DRY LAND BELT COUNTRIES, SUB-SAHARAN AFRICA (SSA), SOUTH AND SOUTHEAST ASIA, LATIN AMERICA, SMALL ISLAND DEVELOPING STATES (SIDS) AND THE ARCTIC

People living in low-lying areas, the semi-arid dryland belt along the Sahel that separates Africa's arid north from extra fertile areas, merely flooded regions on the Equator, and formation regions are presumptively to be affected.

The following countries and regions are thought-about the foremost at risk of climate change:

- The semi-arid object belt countries due to overall vulnerability to droughts from the Sahara/Sahel to the Middle East, (The most affected countries embody Niger, Sudan, Ethiopia, Somalia).
- SSA attributable due to vulnerability to droughts and floods, (The most affected countries embrace the Republic of Kenya, Uganda, Tanzania, Nigeria, Mozambique, and South Africa).

- little island developing states attributable due to water level rise and cyclones, (The most affected countries embody the Federal Islamic Republic of the Comoros islands, Kiribati, Tuvalu, the Maldives, and Haiti.)
- The Arctic region attributable due to the melting of ice caps.

The region at the most exceedingly immediate risk of droughts and floods is SSA. Droughts are most probable in Burkina Faso, Mozambique, Rwanda, Somalia, and Tanzania, whereas the Republic of Malawi, Mozambique, Nigeria, Somalia, Sudan, and Tanzania are thought of particularly at risk of floods. Flooding is to boot in all probability in South Asia (Afghanistan, East Pakistan, and Nepal). The foremost storm-prone areas are on the coasts of East Africa (Mozambique, Madagascar) and South Asia (Bangladesh) equally as in the Southeastern and central areas of the USA. (Watkins, 2007; Ehrhart, 2008), Figure 7.3 shows the area of natural vulnerability to floods, storms, droughts, and sea-level rise.

7.11   CLIMATE CHANGE REDUCES FOOD SECURITY—
ESPECIALLY IN THE POOREST PARTS OF THE WORLD WHERE
HUNGER IS ALREADY AN ISSUE

Weather-related disasters destroy crops and reduce soil quality in a very range of the world's poorest regions. Exaggerated temperatures, shriveled rain, water shortages, and drought reduce yield and stock health. The geological process shows away at the amount of cultivatable land and thus the standard of the soil. Within the world's oceans, international temperature change and reef destruction reduce fish stock. The impacts are notably severe in developing regions like SSA and thus the article belt that stretches across the Sahara Desert and therefore the Middle East all the way to parts of China (Fischer et al., 2005; Parry, 2007; Easterling et al., 2007; Erda et al., 2009).

While hotter temperatures are leading to extra favorable agricultural conditions and exaggerated yield in some parts of North America and Russia, the worldwide impact of temperature change on overall food production is negative (Fischer et al., 2005). The injury is especially severe at intervals the world's poorest areas, where subsistence farmers get hit double by the less favorable growing conditions. First, many farmers may not have enough crop production to feed their families. Second, the inadequacy of their own crop would possibly in all probability force them to buy for food at a time once prices area unit high because of reduced international crop yields and increment. In 2008, the Food and Agriculture Organization (FAO) of the UN derived that quite 900 million are afflicted with hunger, or concerning 13% of the worldwide population (Assumes global population equals 6.76 billion) (Diouf, 2009).

FIGURE 7.3   Area of natural vulnerability to floods, storms, droughts, and sea-level rise (Hamada, 2018).

Of those tormented from hunger, 94% live in 72 board developing nations (World Food Program estimates that 963 million are hungry today; 907 million of that live in developing nations, 565 million are in

Asia and thus the Pacific, 230 million in SSA, 58.4 million in SSA and thus the Caribbean, 41.6 million at the SSA and North Africa, etc., (WFP, 2009). Figure 7.4 shows that desertification agriculture land due to climate change in Upper Egypt. Most are subsistence farmers, landless families, or folks operating in work or biological science. The remainder lives in shanty cities on the fringes of urban areas. A quarter of the hungry populations in the world are children (WFP, 2009). International climate change is projected to be at the root of malnutrition disease for relating to 45 million people, as a results of reduced agricultural yields of cereals, fruits, vegetables, livestock, and dairy, equally as a result of the cash crops like cotton and fish that generate gain. For example, drought hurts crops in Africa where over 90% of farmers are little scale and relating to 65% of people's primary supply of monetary gain is agriculture (IFPRI, 2004).

FIGURE 7.4   Desertification agriculture land due to climate change in Upper Egypt, Egypt (Hamada, 2018).

7.12 BY 2030, THE NUMBER OF HUNGRY PEOPLE BECAUSE OF CLIMATE CHANGE IS EXPECTED TO GROW BY MORE THAN TWO THIRDS

Within 20 years, the amount of hungry people as a result of worldwide temperature change is projected to just about double to 75 million (Parry et al., 2005; Easterling et al., 2007). The explanation for this increase is that the results of worldwide climate change become extra pronounced as temperatures rise at intervals the same 20 years, international climate change is projected to chop back international food production by roughly 50 million tons (Parry et al., 2005). That, in turn, would possibly force up international food prices by 20% for ton, the estimates for production losses and increase are supported crop model projections that focus on changes in yield. Completely different models exist, like agro-economic models, that come back up with somewhat extra conservative estimates, (this model includes assumptions related to mitigating actions by economic actors like shifting crop production between regions, (Parry et al., 2005; Fischer et al., 2005; Easterling et al., 2007). In some parts of Africa, climate change is anticipated to cut back give up to 50% by 2020 (Boko et al., 2007; Parry et al., 2007). Historical proof shows that higher food costs cause an on the spot and direct jump in hunger levels. Throughout the 2008 food crisis, the amount of hungry people at intervals the globe exaggerated by 40 million, primarily because of exaggerated food costs (Diouf, 2009). Figure 7.5 shows global climate change reduces food security.

FIGURE 7.5   Climate change reduces food security (Hamada, 2018).

7.13 POVERTY INCREASES VULNERABILITY TO CLIMATE CHANGE

While all countries are affected by means of climate change, we find that the poorest countries may be extra exposed, as they often have climates in the direction of dangerous physical thresholds. They also rely more on outside work and natural capital and have less economic approach to conform quickly. The risk associated with the effect on workability from growing heat and humidity is one instance of the way poorer countries might be extra at risk of climate hazards. When looking on the workability indicator (that is, the share of effective annual outside working hours lost to extreme heat and humidity), the top quartile of countries (based on GDP per capita) have an average increase in risk by means of 2050 of approximately 1 to 3% points, whereas the bottom quartile faces an average boom in danger of about 5 to 10% points. Lethal heat waves display much less of a correlation with in keeping with capita GDP, however it is critical to note that several of the most affected nations—Bangladesh, India, and Pakistan, to name a few—have relatively low per capita GDP levels.

The CCVI constitutes of six factors: economy; natural resources and ecosystems; impoverishment, development and health; agriculture; population, settlement and infrastructure; and institutions, governance and social capital. A sub-index was developed for each cluster and these were combined to create the CCVI. The natural resources and ecosystems and agriculture sub-indices area unit weighted double as heavily as a result of the others. The index values vary from 0 to 10, where zero equals highest risk and 10 equals lowest risk. For the wants of this analysis, have been made public 'vulnerable' countries as those with a mean CCVI of 5 or less and 'extremely vulnerable' countries as those with a mean CCVI of 2.5 or less, are at risk of international climate change nowadays in socio-economic terms. The worldwide poor, with incomes of however $2 per day (40% of world population), have very restricted resources to reply and adapt to international climate change whereas not facilitate. People with incomes between $2 and $10 have some capability to reply but they are still in all probability to be vulnerable if confronted with the impacts of worldwide climate change. Those relying on natural resources for his or her livelihoods like farmers, fishermen and low-wage earners in business are going to be notably at risk of gain losses because of international climate change. The degree of social development and native infrastructure to boot significantly determines the vulnerability of communities and their capability to adapt. People living whereas not access to low-cost health care, water, electricity and made-up roads area unit extra in all probability, suffer from severe human impact than those that have accessed to those basic services (Friedman, 2008).

Broad lack of access to insurance in developing countries extra magnifies the vulnerabilities. Insurance is also a mean for people to help them achieve their own answer of a crisis and this cowl against risks can help individuals escape impoverishment. Worst affected regions embody the Sahara Desert, the coastline of Eastern Africa, all of South Asia, and plenty of little island states. Africa is that the most vulnerable region—15 of the world's 20 most vulnerable countries are African, in disparity developed nations are the tiniest amount vulnerable—in particular Scandinavia, Canada and thus the US—every due to their lower exposure to the physical impact of the worldwide climate change and thus the larger investment in climate change adaptation, like coastal protection and advanced warning systems (McCarthy, 2008).

Particular attention ought to run to the around 500 million who exist in countries that are terribly prone to climate change due to the physical location of their homes and social circumstances. Figure 7.6 shows water investments in Noubaria at North West Egypt. The terribly vulnerable people are typically poor and live in least developed countries that are in danger of quite one form of weather disaster, as floods, droughts and storms, gradual environmental degradation like water level rise or geologic process. The 10 most vulnerable countries are Comoros, Somalia, Burundi, Yemen, Niger, Eritrea, Afghanistan, Ethiopia, Chad and Rwanda. These 10 most vulnerable nations have full-fledged nearly 180 storms or floods throughout the last 30 years. In these same countries, eleven million were full of drought in 2008 alone whereas 85 million are full of droughts in last 30 years (CRED and ISDR (2008).

Adaptation can reduce overall vulnerability, in particular among the worlds poorest. This might be through policies of finance in early warning and evacuation systems to arrange people for storms, or serving to farmers to change the crops grownup and thus the temporal property of planting and gather. The great news is that there are some success stories of poor countries reducing vulnerability to the impacts of climate change. Bangladesh, one altogether the country's most naturally prone to climate change, has taken steps over the past few years to become higher prepared, and thus, less vulnerable. These steps helped reduce mortality in People's Republic of Bangladesh throughout Cyclone Sidr in 2007 that killed around 40 times fewer people than the identical scale cyclone in 1991 (3,400 deaths versus 138,000) that is despite the following population will increase over the intervening period.

FIGURE 7.6   Water investments, Noubaria; North West Egypt, Egypt (Hamada, 2018).

7.14   THE POOREST ARE HARDEST HIT BUT CLIMATE CHANGE IS A GLOBAL PROBLEM

Even though over 90% of all weather-related disasters happen in developing countries, developed nations are also affected—a lot of and a lot of with devastating effects. The human impact of recent heat waves, floods, storms, and forest fires in Europe, the US and Australia is shocking. The

2003 wave in Europe killed 35,000 people (Epstein, 2006) and cyclone Katrina that hit the US coast in 2005 caused economic losses in far more than USD 100 billion. California's $35 billion farm trade, that's the 1/2 all US fruit, vegetables, and nut production, is extremely at risk of climate change—significantly to the drought that reduces the water supply needed to grow crops (Milliken, 2009).

The American state is exposed to extreme coastal storms that might have a sway on some 480,000 people and cause injury to homes, businesses, power plants, ports, and airports derived at over $100 billion over the consecutive year (Spotts, 2009). Australia is probably the developed country most prone to the direct impacts of worldwide climate change and the indirect impact from neighboring countries that are stressed by climate change (Garnaut et al., 2008). Figure 7.7 shows water, agriculture, and growth in Upper Egypt. The temperature has exaggerated by three-quarters of a C degree in the past 15 years in Australia and rain has

FIGURE 7.7   Water, agriculture, and growth, Upper Egypt, Egypt (Hamada, 2018).

reduced—leading to water deficiency and drought (Draper, 2009). The multi-year drought since 2001 in South-eastern Australia is that the worst at intervals the country's recorded history (Draper, 2009). It's calculable that gross domestic product (GDP) was reduced by 1% in 2002–2003 as a result, claiming 100,000 jobs. In 2003, grain output minimized by 50%, numerous sheep and cows died and over 80% of dairy farmers were compact (TWM, 2009). Partially, the identical multi-year drought, wheat costs in Australia jumped 42% from 2007 to 2008 (Morrison et al., 2009).

7.15 DEVELOPMENT GOALS AND HUMANITARIAN RELIEF AT RISK

Climate change significantly impacts the international community's development facilitate and humanitarian relief efforts. The human impact of climate change is anticipated to have a real worth every in terms of lost progress towards development goals and exaggerated prices of facilitate.

7.16 CLIMATE CHANGE THREATENS SUSTAINABLE DEVELOPMENT, ESPECIALLY THE MILLENNIUM DEVELOPMENT GOALS (MDGS)

Climate change slows—and at the worst cases reverses—progress created in fighting destitution and diseases. Unless adaptation to climate change is funded through more channels, the growing impact of climate change is predicted to consume an increasing share of development aid. In fact, the OECD estimates that most of one-half of development assistance is exposed to climate risks (Mitchell and van Aalst, 2008). Official development facilitate alone amounted to $120 billion in 2008. This amount is already lean to reach international development goals and thus the exposure to climate risks extra threatens its value. The progressive risks associated with climate change increase the costs of achieving flagship humans' development goals like reduced childhood mortality and improved nutrition. Climate change is additionally notably damaging to development facilities for the world's most vulnerable groups and communities. Climate change poses a threat to all or any of the eight MDGs. For example, the impact of climate change on poorness, access to natural resources like water, and diseases like protozoan infection have direct implications for the action of many of the MDGs. These goals represent the commitment of the international community to reducing extreme poorness. They were adopted because the United Nations (UN) millennium declaration in September 2000 by world organizations member states and leading development institutions. The declaration launched a series of time sure targets—the MDGs with a deadline of 2015—like halving the quantity of world hunger and poorness. Figure 7.8 shows the desertification in Upper Egypt.

FIGURE 7.8 Desertification show in Upper Egypt, Egypt (Hamada, 2018).

7.17 ADAPTATION IS ALREADY CONSIDERED A VITAL PART OF ANY FUTURE CLIMATE CHANGE

Developing countries are already suffering from the impacts of worldwide climate change and are the foremost prone to future change. Assortment of developing countries has developed adaptation plans or is at intervals the strategy of finalizing them. This includes the National Adaptation Programs of Action (NAPA) of LDCs. There is presently urgency for developing countries to hunt out ways in which to implement these plans. Against a background of low human and cash capability, developing countries lack many of the resources to undertake and try this on their own. Adaptation is already thought of a major part of any future climate change regime. Among the UNFCCC and thus the international community, deliberations are building to hunt out associate degree economical implies that to tackle climate change, that's delineated by international organization executive Ban Ki-moon as a result of the defining issue of our era. Future decisions among the UNFCCC negotiating method ought to assist developing countries in an extremely efficient, innovative, and clearly suggests that, with the transfer of information, technology, and cash resources to adapt and to adapt in any respect levels and altogether sectors (UNFCCC, 2007).

At a series of workshops for Africa, Asia, and Latin America and an expert meeting for Small Island Developing States (SIDS) throughout 2006–2007, these regions acknowledged are as for future action in adapting to climate change. To be handiest, adaptation plans and techniques should be integrated into property development planning with and risk reduction coming up with at community, local, national, and international levels. Crucially there has been little work to integrate adaptation into development plans or among existing poorness alleviation frameworks. Taking stock of and promoting wise observe by the international community at intervals the mixing of climate change connected issues would facilitate promote adaptation strategies with many advantages. Capability continues to be needed to change developing countries to develop adaptation programs and techniques. The countries' work program is building capability to understand and assess impacts, vulnerability, and adaptation and to make suggested decisions on wise adaptation actions and measures. The NAPAs have proved a really vital because of priorities adaptation actions for LDCs. Initiating a technique for extending the positive experience of NAPAs for developing countries, that do not appear to be LDCs, which, wish to develop national adaptation programs or strategies, could vitally facilitate adaptation alternative prioritization. This would possibly take into thought lessons learned from the petsai preparation methodology and its victorious experience at policy integration, to boot as relevant outcomes from the country work program. Mistreatment native cope methods can assist community-based adaptation and could be fast by information exchange among fully completely different communities facing similar problems, like via the UNFCCC native cope methods information. Finding synergies between the cities conventions could boot facilitate share information and information on assessment processes. If there are delays in implementing adaptation in developing countries, along with delays in finance adaptation comes, this might lead ultimately to exaggerated prices. Delays in implementing adaptation conjointly can lead to larger dangers to lots of people. For example, extreme events along with droughts, floods, and loss of glacial soften water could trigger large-scale population movements and large-scale conflict due to competition over scarcer resources like water, food, and energy (UNFCCC, 2007).

There are already mechanisms for cash facilitate for developing countries available. Application procedures should be efficient, along with enhancing the aptitude for the event of project proposals to boot as capacity-building to identify the assorted desires and modalities of assorted sources of current support. It's to boot clear that current funding is not enough to support adaptation needs. Recent studies by the UNFCCC secretariat showed that a progressive level of annual investment and cash flows of regarding USD 50 billion is needed for adaptation in 2030. In the context of any discussion on future international cooperation on climate change, future cash resources should be enough, inevitable, and property therefore on facilitate adaptation to the adverse impacts of climate change by developing countries. To boot as via funding envisioned through the operationalization of the distinction Fund beneath the city Protocol, innovative finance decisions are needed to shut the gap between costs of adaptation and available resources. Insurance could be a section that has been called a really vital part of future action on adaptation. Innovative risk-sharing mechanisms are needed to retort to the new challenges exhibit, along with the increasing frequency of most events, land degradation, and loss of diversity. Collaboration and cooperation between South-South and North-South can directly have interaction, multiple stakeholders, in addressing climate change and coordinate coming up with and actions. This could be fast by international fore with the participation of Parties and relevant stakeholders involved in South-South and North-South collaboration, many-sided environmental agreements and with the disaster risk reduction community. Awareness-raising among the key sectors and mass media, along with victimization current events like and health crisis could facilitate promote adaptation measures with co-benefits. Action desires a world framework for international cooperation. Adaptation with climate change is also a really vital part of this framework. Actions to modify adaptation to climate change produce opportunities to property development. Developing countries, need resources therefore on promote these actions? A victorious framework ought to directly involve facilitate for adaptation in developing countries, notably very little island developing States and LDCs, provided that they go to disproportionately bear the force of climate change impacts (UNFCCC, 2007).

7.18   WHAT CAN WE EXPECT IN THE FUTURE?

Agriculture is usually the first trade to point the results of drought through low crop yields and dugout levels. Droughts that persist over 2 or 3 years begin to eat up groundwater supplies; lower stream, and lake levels; and reduce runoff in major watersheds. Once major bodies of water are affected, municipalities, and completely different industries like recreation and business, fisheries, and transportation ought to conjointly cope with the results of drought. The variability of climate and weather conditions is also a key to give some thought to agricultural production. These conditions modification from year-to-year and are projected to vary even lots of in the long run than they're doing presently. Many scientists believe that climate is undergoing very important international modification, with an increase in any regional variability. Adaptation to climate and weather risk is implicit in the continuing development of the agro-food sector. Agricultural systems have evolved to deal with some variations in climate, but are additionally prone to extremes like high winds, excessive or lean precipitation, hail, or extreme temperatures (Tremblay, 2010). This updated Southeast Mediterranean Sea zone builds upon the initial key methods to raised manage the risks of drought and climate change throughout three made public situations: 1) ancient or near ancient conditions, 2) exceptional/notable conditions; and 3) extreme conditions. Every farm desires minimum amounts of rain at varied times to continue functioning usually. One thing however the minimum will mean that associate degree modification in coming up with and decision-making is required. The crucial amount and temporal property will depend upon hold on soil condition reserves, the actual crop, pasture, or livestock. An action founded is usually established which can embrace response and/or decision times. Designing for drought could be a dynamic method that evolves through trial and error. As we have a tendency to tend to still study the regional and continental scale of weather and climate dynamics and its result on the water and energy cycle over Southeast Mediterranean Sea zone will adapt. Coming up with for drought is additionally joined to coming up with for climate change as a result of the exaggerated risk of drought for Southeast Mediterranean Sea zone producers is maybe going as climate change. Drought and climate change do not appear to be problems to be solved; rather, they are risks that must be managed.

7.19 FARM-LEVEL RISKS AND CONSTRAINTS

Risks faced by farmers are many and varied, and are specific to the country, climate, and native agricultural production systems. These risks and their impacts on farmers are widely classified in the literature and thus we'll not get to cover the issues here (Barnett and Coble, 2008). To boot, farmers face constraints that do not alter them to either improve or increase their production and revenues. Samples of such constraints are restricted access to finance, dislocation from markets, poor access to inputs, lack of conductive services and information, and poor infrastructure (for example, irrigation or rural roads). These constraints are typically worse in low-income countries, wherever public merchandise and personal sector service delivery are usually poorly developed. The importance of noting the excellence between a risk and a constraint is that typically the latter is performed of the previous. For example, many argue (and it would seem logical) that access to finance (in terms of every worth and availability) for farmers in developing countries would improve if the potential financiers were able to be assured that the risks inherent with agricultural production had been managed, thereby reducing their compensation risk. Of course, many constraints are usually not driven by one underlying risk alone.

Taking access to finance everywhere once more as Associate of example, albeit the underlying weather risk is managed through the acquisition of Associate in insurance product or installation of irrigation, this still leaves the financier running form of risks. For example, the farmer may just sell the merchandise and not repay the bank, or costs may fall to such Associate in insurance extent at harvest that the revenue is lean to repay the loan amount, or even the crop was destroyed by locusts and there was no crop left to sell at the due compensation date. The difficulty of the existence of multiple risks in agriculture should be noted. Only too usually, the apparent management of one major risk leaves stakeholders (although rarely the farmers) with the impression that the final risk profile has been managed. However, this may be usually not the case; even once farmers and their partners have managed their own direct risks, indirect risks can cause losses.

For example, a pandemic of afflation in maize in a given country may lead to the imposition of an import ban by potential patrons. Albeit farmers and thus the provide chain they are concerned may have well managed this risk and their maize is afflation free, they go to suffer from the country's market access restrictions. Equally, whether or not a farmer has managed contamination risks in his own basket of product, should the method, or fail to manage its crop assortment or process activities properly, then the farmer would possibly suffer attribute able to the exclusion of the processor from the market (there being no completely different vendee for the farmer's produce). Therefore, thought of risk throughout a provide chain permits lots of comprehensive assessment and management of risks (Jaffee et al., 2010).

7.20   WEATHER RISKS IN AGRICULTURE

Obviously, given the massive selection and complexities of world climates, it's powerful to generalize once discussing weather-related risks. The impacts of a given weather event dissent to keep with the actual agri-cultural system, variable water balances, form of soil and crop, and avail-ability of various risk management tools, (such as irrigation). To boot, the negative impacts of weather events are usually aggravated by poor infra-structure (such as poor drainage) and management. From a weather risk management point of view, there are two main sorts of risks to give some thought to. These relate to (1) sudden, unforeseen events (for example, windstorms or important rain) (2) accumulative events that occur over an extended amount (for example, drought). The impacts that either of these sorts of risk have to vary wide consistent with crop kind and choice and temporal order arrangement of occurrences (Jaffee et al., 2010).

7.21 RELATIONSHIP BETWEEN WEATHER AND YIELDS

Short-duration extreme weather events (such as hail, windstorm, or frost) can cause devastating direct injury to crops at intervals the fields. Assessment of these damages is usually beneath taken without delay by examination. On the opposite hand, whereas the final word outcome of accumulative events is usually devastatingly obvious, lots of the injury already occurred earlier throughout a stage of crop development. However, correlations of the weather event and injury are usually powerful to model, other than the foremost extreme events. At intervals, in the case of accumulative rain deficit (drought), the foremost effective correlations exist for rain-fed crops big up in areas where there is a clear sensitivity of the crop to deficits in available water, and clearly made public rainy seasons. An example is maize production in southern African countries, like the Republic Zambia and the Republic of Malawi. Less-clear relationships are found in areas of upper and a lot of regular precipitation or less-clear seasonality, or where completely different influences, like pest and malady, are very important causes of crop losses. If partial or full irrigation is in place, the relationships become lots of less durable. Rain-fed production at the tropics (where precipitation is higher and fewer seasonally marked) is an example where correlations are additionally less undemanding to see. However, droughts are also a feature of tropical crop production, anywhere both floods and droughts can occur at the identical year. In sum, creating generalizations is risky (Jaffee et al., 2010).

7.22 WEATHER RISK MANAGEMENT APPROACHES

Risk mitigation, coping, and transfer are the key strategies in agricultural risk management. These strategies are usually applied at the family, community, and market levels (Jaffee et al., 2010). The key points are:

- Agriculture is concerning to several sorts of risks that expose farmers, business entities, and governments to potential losses.
- Several approaches are utilized to manage agriculture-related risks, and sometimes several should be applied among an overall risk management framework.

- Risk management methods enclose risk mitigation, risk transfer, and risk copes.
- Formal (market-based) approaches (including agricultural finance and insurance) allow disciplined cash management of risks but are usually tough to implement in developing countries and cannot be acceptable for managing extreme risks or disasters.
- Informal approaches are rather a lot of usually found at the farmer level in developing countries. They embrace savings, family buffer stocks, community savings, and no formalized mutual.

7.23   ADAPTATION OPTIONS

So far, solely restricted adaptation measures considering climate change have been undertaken, in every developed and developing countries. These adaptation measures undertaken by sectors are victimization fully completely different technologies to forestall environmental damages (Adger et al., 2007). In many cases, people adapt to action by driving their behaviors, driving their occupation, by moving to a fully different location, or usually they're going to use fully different types of technologies. These can either be hard technologies; like solely restricted new irrigation systems, drought-resistant crop varieties, newly introduced insect and pesterer resistance varieties and mistreatment of the results of different new breeding techniques. For example, mistreatment product from techniques of plant biotechnology and genetic modification (a technique that discovers specific sequence, but they work, associate degreed establish traits and transfer sequence where they are needed), or soft technologies; as an example, insurances, completely different services, and crop rotation or they may use the mixture of laborious and soft (Stalker, 2006).

7.24   ADAPTATION STRATEGIES FOR AGRICULTURE

Agriculture is very sensitive to even minor climate variations, and have a bearing on agricultural output even for one season, so current action can have a sway on long-term agricultural productivity and food security (Stalker, 2006). Climate change impacts on crops as have a way on human health, largely through potential for malnutrition and as a result, few studies have calculable numerous people are in risk of hunger (Warren, 2011).

Adaptation is economical if value of making efforts could be a smaller quantity than the following advantages: joint variations are beneficiary given that it's through governmental actions, additionally, political forces perhaps impressed if governments are engaged in inefficient adaptation behavior, thus, it is not in any estimate clear whether or not economical levels of joint adaptation are undertaken (Mendelsohn, 2000).

7.25 SUSTAINABLE ADAPTATION MEASURES

The physical impacts of worldwide climate change in existing programs and activities should ponder wide vulnerability to climate change and these activities are necessary in adaptation to contribute in poorness reduction. Organization and development agencies should focus on risk reduction works, e.g., early warning and moving of people from danger areas. The risks are variable from place to position and between fully completely different teams; measures targeted at risks are very specific to a selected situation. An example, if we have a tendency to take agricultural productivity reduction caused by climate change stress could doubles targeted to resolve through adaptation measures, which may focus in ever-changing cropping pattern techniques and mistreatment fully completely different trendy technologies (Eriksen and O'Brien, 2007). Since fully completely different outcomes and adaptation, responses in associate degree extremely sure cluster could have a sway on the vulnerability context of another team elsewhere and adaptation responses could have a sway on socio-environmental transformations, property adaptation is so a world environmental issue (Eriksen, 2009). In designing adaptation and decision-making processes, establishments should specialize in low-risk information, on capability building measures and specific actions to cut back vulnerability, and conjointly the crucial thought is not entirely coming up with but also the implementation of the different processes.

7.26 CONCLUSION

This chapter gives a quick evaluate of history of Future threats to agricultural food production, wherein global food security is one of the maximum pressing issues for humanity, and agricultural production is crucial for attaining this. The existing analyses of specific threats to agricultural food production seldom carry out the contrasts associated with one of kind tiers of financial improvement and different climatic zones. So by way of investigated the equal biophysical threats in three modeled kinds of international locations with different financial and climatic conditions. The threats analyzed had been environmental degradation, weather change and diseases and pests of animals and plants. These threats have been analyzed with a methodology permitting the associated risks to be compared. The world population will have increased to 9 billion people, there will be a larger middle class within the world and climate change will be causing more extreme weather events, higher temperatures and changed precipitation. It is suggested that the risks, presented by using the biophysical threats analyzed, differ among the three modeled types of countries and that weather zone, public stewardship and economic strength are predominant determinants of these differences. These determinants are far from evenly spread among the international's most important food producers, which implies that diversification of risk tracking and international assessment of agricultural production is vital for as-Suring global food security in 2050.

KEYWORDS

- climate change
- fertile soils
- glaciers
- greenhouse gasses
- millennium development goals (Mdgs)
- world emissions

REFERENCES

Adger, W. N., Agrawala, S., Mirza, M. M. Q., Conde, C., O'Brien, K., Pulhin, J., Pulwarty, R., Smitand, B., & Takahashi, K., (2007). Assessment of adaptation practices, options, constraints and capacity. In: Parry, M. L., Canziani, O. F., Palutikof, J. P., Van der Linden, P.J., & Hanson, C. E., (eds.), *Climate Change 2007: Impacts, Adaptation and Vulnerability: Contribution of Working Group II to the Fourth Assessment Report of* the *Intergovernmental Panel on Climate Change* (pp. 717–743). Cambridge University Press, Cambridge, UK.

Allan, R., & Soden, B. (2008). Atmospheric warming and the amplification of precipitation extremes. *Science 321*(5895), 1481–1484. doi:10.1126/science.1160787.

André, T., (2010). *Agricultural Drought Program History in Alberta*. Bob Barss ADMC Advisory Group Representative, ADRMP.

Annan, K., (2009). *The Anatomy of a Silent Crisis, Global Humanitarian Forum*. Global humanitarian forum; chair, steering group, human impact report: climate change.

Barnett, B., & Coble, K., (2008). Poverty traps and index-based risk transfer products. *World Development, 36*, 1766–1785.

Boko, M., et al., (2007). *African Climate Change, Impacts, Adaptations, and Vulnerability*. Contribution of working group II to the fourth assessment report of the Intergovernmental Panel on Climate Change, M.L.

Burke, J. J., & Moench, M., (2000). Groundwater and society, resources, tensions and opportunities. In: *Themes in Groundwater Management for the 21st Century* (p. 170). United Nations, New York.

Cai, X., Ringler, C., & Rosegrant, M. W., (2001). *Does Efficient Water Management Matter?: Physical and Economic Efficiency of Water Use in the River Basin*. EPTD discussion papers 72, International Food Policy Research Institute (IFPRI).

CRED, ISDR, (2008). Disasters in numbers. *International Strategy for Disaster Reduction and Centre for Research on the Epidemiology of Disasters*, pp. 1, 2.

Diouf, J., (2009). *Food Security for All*. Food and Agriculture Organization of the United Nations high-level meeting, Madrid. http://www.fao.org/english/dg/2009/2627january2009.html (accessed on 4 January 2020).

Draper, R. (2009). Australia's Dry Run, National Geographic Magazine, *215*(4), 34–59.

Easterling, W. E., et al., (2007). Food, fiber, and forest products. In: Parry, M. L., et al., (eds.), *Climate Change 2007: Impacts, Adaptations and Vulnerability: Contribution of Working Group II to the Fourth Assessment Report of the Intergovernmental Panel on Climate Change* (pp. 273–313). Cambridge University Press, Cambridge, UK.

Ehrhart, C., (2008). *Humanitarian Implications of Climate Change Mapping Emerging Trends and Risk Hotspots* (p. 2). CARE.

Emerton, L., & Bos, E., (2004). *Value Counting Ecosystems as an Economic Part of Water Infrastructure* (p. 88). IUCN, Gland, Switzerland and Cambridge, UK.

Epstein, P. R., (2006). *Climate Change Futures: Health, Ecological and Economic Dimensions*. The Center for Health and the Global Environment, Harvard Medical School.

Erda, L., Hui, J., Wei, X., et al. (2009). "*Climate Change and Food Security in China.*" The Chinese Academy of Agricultural Science and Greenpeace China, 69, Beijing: Xueyuan Press, 2008.

Eriksen, S. H., & O'Brien, K., (2007). Vulnerability, poverty and the need for sustainable adaptation measures. *Climate Policy, 7*, 337–352.

Eriksen, S., (2009). *Sustainable Adaptation Emphasizing Local and Global Equity and Environmental Integrity*. IHDP.

FAO, (2004). Economic valuation of water resources in agriculture: From a sectorial to a functional perspective of natural resource management. *FAO Water Report, 27*, 186.

Fischer, G., et al., (2005). Socio-economic and climate change impacts on agriculture: An integrated assessment, 1990–2080. *Philosophical Transactions of the Royal Society, 360*, 2067–2083.

Friedman, T. L., (2008). *Hot, Flat and Crowded: Why We Need a Green Revolution and How it Can Renew America* (p. 158). Farrar, Straus and Giroux.

Garnaut, R., et al., (2008). *The Garnaut Climate Change Review.* Cambridge University Press. http://www.garnautreview.org.au/index.htm#pdf (accessed on 4 January 2020).

Hamada, Y. M. (2018). Special pictures album, Took by Youssef M. Hamada, Egypt, 2018.

Hussain, I., (2005). *Pro-Poor Intervention Strategies in Irrigated Agriculture in Asia.* Poverty in irrigated agriculture: Issues, lessons, options and guidelines, final synthesis report submitted to the Asian Development Bank, International Water Management Institute (IWMI), Colombo, Sri Lanka.

IFPRI, (2004). *Ending Hunger in Africa Prospects for the Small Farmer.* International Food Policy Research Institute. http://www.ifpri.org/pubs/ib/ib16.pdf (accessed on 16 January 2020).

ISDRS (2008). International Strategy for Disaster Reduction Statistics, ISDR secretariat Biennial Work plan, Final Report. 2008–2009, www.preventionweb.net, 2008–2009.

Jacob, B., & Jean-Marc, F., (2011). *Climate Change, Water and Food Security: Land and Water Division.* Food and Agriculture Organization, United Nations, Rome.

Jaffee, S., Siegel. P., & Andrews, C., (2010). *Rapid Agricultural Supply Chain Risk Assessment: A Conceptual Framework.* The World Bank, Washington. D.C.

Kundzewicz, Z. W., et al., (2007). Freshwater resources and their management. In: Parry, M. L., et al., (eds.), *Climate Change 2007: Impacts, Adaptation, and Vulnerability: Contribution of Working Group II to the Fourth Assessment Report of the Intergovernmental Panel on Climate Change* (pp. 173–210). Cambridge University Press, Cambridge, UK.

Margareta Wahlström (2009). *Human Impact of Climate Change*; Global Humanitarian Forum – Human Impact Report: Climate Change – Geneva © 2009.

McCarthy, M., (2008). *Why Canada is the Best Haven From Climate Change.* UK Independent. http://www.independent.co.uk/environment/climatechange/why-canada-is-the-best-haven-from-climatechange-860001.html (accessed on 4 January 2020).

McMichael, A. J., et al., (2004). *Chapter 20: Global Climate Change.* In comparative quantification of health risks, the World Health Organization.

Mendelsohn, R., (2000). *Efficient Adaptation to Climate Change* (Vol. 45, pp. 583–600). Kluwer Academic Publishers, Netherlands.

Milliken, M., (2009). *Water Scarcity Clouds California Farming's Future.* Reuters. http://www.reuters.com/article/domesticNews/idUSTRE52C07R20090313?sp=true (accessed on 4 January 2020).

Mitchell, T., & Van Aalst, M., (2008). *Convergence of Disaster Risk Reduction and Climate Change Adaptation.* DFID. http://www.research4development.info/PDF/Articles/Convergence_of_DRR_and_CCA.pdf (accessed on 4 January 2020).

Morrison, J., et al., (2009). *Water Scarcity and Climate Change: Growing Risks for Businesses and Investors* (p. 6). Ceres and the Pacific Institute. http://www.ceres.org/Document.Doc?id=406 (accessed on 4 January 2020).

MRS (2009). Munich Re Statistics, Munich Re achieves a strong result, Risks Climate Change, One of humanity's greatest challenges, https://www.munichre.com/en.html.

Parry, et al., (2007). *African Climate Change, Impacts: Adaptations and Vulnerability* (pp. 433–467). Cambridge University Press, Cambridge, UK.

Parry, M. L., (2007). *The Impacts of Climate Change for Crop Yields: Global Food Supply and Risk of Hunger* (Vol. 4, No. 1, p. 12). ICRISAT. http://www.icrisat.cgiar.org/Journal/SpecialProject/sp14.pdf (accessed on 4 January 2020).

Parry, M., Rosen, Z. C., & Livermore, M., (2005). *Climate Change: Global Food Supply and Risk of Hunger* (Vol. 360, pp. 2125–2138). Philosophical Transactions of the Royal Society.

Sperling, F., (2003). *Poverty and Climate Change: Reducing the Vulnerability of the Poor Through Adaptation.* Washington, DC: AfDB, AsDB, DFID, Netherlands, EC, Germany, OECD, UNDP, UNEP and the World Bank (VARG).

Spotts, P., (2009). *California's Climate Change Bill Could Top $100 Billion.* The Christian Science Monitor. http://features.csmonitor.com/environment/2009/03/11/california%E2%80%99sclimate-change-bill-could-top-100-billion/ (accessed on 4 January 2020).

Stalker, P., (2006). *Technologies for Adaptation to Climate Change (UNFCCC).* Bonn, Germany.

Tin, T., (2007). *Climate Change: Faster, Sooner, Stronger.* An overview of the climate science published since the 4th IPCC assessment report. World Wildlife Foundation, WWF and Allan, R., & Soden, B., (2008). Atmospheric warming and the amplification of precipitation extremes. *Science, 32*(5895). doi: 10.1126/science.1160787.

TWM, (2009). *Drought Australia, 2003–2009.* TWM. http://twm.co.nz/ausdrght.htm (accessed on 4 January 2020).

UNEP, (2009). *United Nations Environment Program.* Global Resource Information Database Geneva.

UNFCCC, (2007). *Climate Change: Impacts: Vulnerabilities and Adaptation in Developing Countries.* United Nations Framework Convention on Climate Change. www.unfccc.int (accessed on 4 January 2020).

Walter Fust, (2009). *Global Humanitarian Forum.* Chair steering group, human impact report: Climate change, CEO/Director-General.

Warren, R., (2011). The role of interactions in a world implementing adaptation and mitigation solutions to climate change. *Phil. Trans. R. Soc. A, 369*, 217–241.

Watkins, K., (2007). *Human Develop Report 2007/2008 Fighting Climate Change: Human Solidarity in a Divided World.* United Nations Development Programme.

WFP, (2009). *Who Are the Hungry?* World Food Program. http://www.wfp.org/hunger/who-are (accessed on 4 January 2020).

WHO, (2004). *The Global Burden of Disease: 2004 Update.* World Health Organization, health statistics and health information systems. http://www.who.int/healthinfo/global_burden_disease/2004_report_update/en/index.html (accessed on 4 January 2020).

CHAPTER 8

Promoting Agribusiness Development in Southeast Mediterranean Sea

ABSTRACT

Climate change is anticipated to decrease the creating potential of the next generation as a result of it cuts family gain and can increase the number of starving kids. Economists guess that every child whose bodily and mental development is inferior by hunger and disease stands to lose 5 to 10% in life earnings (WFP, 2009). As incomes drop, poor families may be forced to send their kids to figure to gain any revenue. Consequently, global climate change affects instructive opportunities and thereby gains potential of consecutive generation. The majority of people are suffering from the impacts of global climate change that are previously terribly poor.

Extreme poverty remains stubbornly excessive in low-income nations and countries affected by conflict and political upheaval, specifically in sub-Saharan Africa. Among the 736 million those who lived on less than $1.90 a day in 2015, 413 million had been in sub-Saharan Africa. This figure has been climbing in latest years and is higher than the number of poor humans in the rest of the world combined. Where forecasts propose that without significant shifts in policy, excessive poverty will nevertheless be within the double digits in sub-Saharan Africa by 2030. About 79% of the world's poor stays in rural areas. The poverty rate in rural areas is 17.2%—extra than 3 times higher than in urban areas (5.3%). Close to half (46%) of extremely poor human beings are youngsters beneath 14 years of age. Where the World Bank is the principle source for global data on intense poverty today and it sets the 'International Poverty Line.' The poverty line revised in 2015—on the grounds that then, a person is taken into consideration to be in intense poverty if they stay on less than 1.90 international dollars (int.-$) per day. This poverty dimension is based on the economic cost of a person's intake. Income measures, on the alternative hand, are only used for countries in which reliable consumption measures aren't available.

Poverty is a major underlying driving force of disaster risk, so it comes as no surprise that the poorest countries are experiencing a disproportionate proportion of damage and loss of existence attributed to disasters. More than 90% of across the world mentioned deaths because of catastrophe arise in low- and middle-earnings countries. Disasters kill 130 people for every one million people beings in low-income countries in comparison to 18 per one million in excessive-income countries. Economic losses as a result from disasters are also much higher in poorer nations, whilst measured as a percentage of their gross home product (GDP). Among the 10 worst disasters in terms of economic damage (whilst expressed relative to GDP), 8 is occurred in low- or middle-income nations.

By 2030, over 20 million less people would be in impoverishment in world with global climate change, presumptuous poverty level of $2 on an each day. This figure is additionally a real under statement as a result of it, among various things, does not account for future increase projections. Typically this is often primarily due to reduced economic gain from lower crop yields with increased food prices. In some countries, the implications of widespread impoverishment from such a cycle are considerably worrisome. In Ethiopia, the globe Bank estimates that water variability would possibly raise impoverishment by 25% (Stern et al., 2006). This chapter displays the problem of promoting agribusiness development in Southeast Mediterranean sea as the views of the most important of specialists in promoting agribusiness development in Southeast Mediterranean.

8.1 BACKGROUND

Agribusiness and agro-industry have the potential to contribute to a ramification of economic and social development processes, likewise as inflated employment generation (particularly feminine employment), gain generation, impoverishment reduction, and enhancements in nutrition, health, and overall food security. Yet, there keep substantial barriers to entirely developing the agribusiness potential evident across Southeast Mediterranean. Figure 8.1 shows Agribusiness as a future of agriculture in the Deshna Sugarcane factory in Egypt. Many of the enabling conditions required for sustainable agribusiness development aren't specific to the arena (or to producing in general), but apply to any or all sectors of the economy. These embrace a stable political economic climate, wise public governance likewise as functioning regulatory institutions, enforceable business laws and property rights, and adequate infrastructure and basic services, likewise as transport, ICTs, and utilities (Yumkella et al., 2011).

FIGURE 8.1   Agribusiness as a future of agriculture, Deshna Sugarcane factory, Egypt (Hamada, 2018).

Historically, they require in addition enclosed the existence of a relatively high-capacity, an interventionist state with active resource

allocation and demand management strategies. A ramification of policies, institutions, and services are further directly relevant to the agribusiness sector. These involve alia: building the desired industrial capabilities
and capacities; upgrading technology and innovation in terms of product
and processes; strengthening group action capacities inside the sector of
production efficiency and business linkages, and cross-border cooperation;
building capability to interchange agro-industrial products; participating
in world, regional, and native price chains; up rural infrastructure and
energy security; promulgating standardization and control measures, and
establishing associated enfranchisement bodies; promoting institutional
services for business; and mobilizing public-private sector cooperation on
business development. Increasing the size and fight of the Southeast Mediterranean Sea's business sector is critical—for farmers, agro-industrial
enterprises, and industry-related services. Indeed, the key challenge for
developing agribusiness in Southeast Mediterranean is that the upgrading
and improvement of manufacturing capacities to beat constraints related
to the event of economical industrial enterprises capable of competitive
in international, regional, and domestic markets (Yumkella et al., 2011).

Driven by globalization and economies of scale, the international
market place for business merchandise is often characterized as market
with some powerful actors—mainly large multinationals and retailers—
seeking the foremost efficient suppliers worldwide, and where cut-throat
competition is prevailing. This has crystal rectifier to a growing concentration, with food method firm's integration backward towards agriculture and
forward towards the retail sector, by ancient markets where tiny holders
sell to native markets and traders. The international market is incredibly
competitive in terms of value and merchandise quality, requiring ICT property and sometimes 'just-in-time' delivery, with resultant high requirements in terms of provision efficiency. Exacting requirements are placed
on suppliers to satisfy conformity standards and specifications, demanded
by shoppers in developed countries and more and more by the growing
category in rising economies. This makes it hard, though unfeasible, for
African agro-industries to 'break-in and move up' the globe price chain
(UNIDO, 2009). In distinction, the national, regional, and sub-regional
markets in Southeast Mediterranean have many competitive advantages
for Southeast Mediterranean producers in terms of proximity of markets
and similarity of shopper preferences. Trendy supermarkets and stores
are quickly increasing in many Southeast Mediterranean countries. A key

challenge is that the involvement of very little farmers within the agro-industry offer chain. Major world players in the business are large international firms operational through a network of subsidiaries and cooperating partners worldwide in varied components of the price chain, like Cargill, prizefighter Alfred Dreyfus Commodities, ADM, Bunge, Wilmer, and Oland. They are giant customers and suppliers of agricultural commodities, employing several thousands of employees internationally. These corporations tend to run their operations by maintaining shut contact with farmers, increasing down the provision chain from the large-volume, thin-profit business of transaction bulk agricultural commodities, to transform some raw materials into premium merchandise to sell at a premium. Whereas typically targeted in certain commodities, the foremost necessary have heterogeneous operations into areas like fertilizers, any as transport and storage, and financial functions (e.g., hedging against associated risks) (Blas and Meyer, 2010).

8.2 SCOPE OF THE CHAPTER

The pace of modification in agribusiness, agro-food, and agro-industrial markets around the world is fast quickly. If Southeast Mediterranean is to exploit these changes, Southeast Mediterranean agro-industry has to bear a structural transformation. In fact, in terms of the transformation of the economy structural modification in farming and agro-industry are tightly reticulate and cannot be analyzed severally of one another. Transformations of the full agribusiness sector involve increasing the productivity of activities at each stage of the varied agriculture-based price chains, whereas at the identical time up coordination among those stages. Improved vertical coordination is significant to achieve the timely flow of productivity-enhancing inputs to farmers and of quality agricultural raw materials to agro-industry. At the identical time, production ought to be closely aligned with the quickly evolving demands of customers (Yumkella et al., 2011).

This chapter examines the supply-side challenges of agribusiness development in Southeast Mediterranean, characteristic the sources of growth through value addition, any as a result of the prime movers for transforming the agro-food system, The term agro-food system is utilized here as shorthand for agriculture and connected agro-industries. It encompasses the interlinked set of activities that run from 'seed to table,' likewise as agricultural input production and distribution, farm-level production, raw product assembly, and method commercialism. It thus encompasses the price chains for numerous agricultural and food merchandise and inputs and conjointly the linkages among them. Whereas it pertains expressly to food, many of the conclusions apply equally well to those components of agriculture and agro-industry that end upon-food merchandise like fibers and biofuels. It concludes by outlining three key ways Southeast Mediterranean governments and their development partners should pursue to beat the constraints of agribusiness development.

8.3   CLIMATE CHANGE ISSUE IN DEVELOPING COUNTRIES

Climate options a significant impact on well-being and levels of happiness. Rehdanz and Madison (2005) show that: Climate changes profit high latitude countries whereas they negatively have a bearing on low latitude countries. Indeed, a touch amount of worldwide warming would increase the happiness of those living in Northern countries, whereas it is the reverse for people living in heat regions. Per Stern (2007) predictions for developing countries reveal alarming future agricultural output and a discount in crop yields, food security and issues related to water. Global climate change involves droughts that are guilty for an increase in food costs, disease, and consequently an increase in health expenditure. Moreover, populations have to influence the issue of water, the foremost climate sensitive economic resource for these countries. In South Asia, for example, global climate change will increase rain and flooding with a straight away impact on agricultural production, and with serious consequences in a very region with a high increase. In Latin America and Caribbean areas, serious threats exist to the rainforests with direct consequences for the subsistence of populations counting on the Amazonian forest. In Sub-Saharan Africa (SSA), an increase in water level threatens coastal cities once higher temperatures raise risks of malnutrition, starvation and malaria, weakened stream flow and conjointly the sequent accessibility of water. Inside the Nile River Basin, the Middle East and North African countries, water stress and severe droughts could cause migration and violent conflicts. In Sub-Saharan Africa (SSA), on 80 million people suffering of starvation thanks to environmental factors, seven million migrated to urge relief food (Myers, 2005). The increasing in temperature of 2°C involve associate degree increasing in population suffering from protozoa infection in Africa of 40–80 million people (70-80 million people affected with associate degree increasing of 3–4°C). By 2020, between 75 and 250 million of African people are going to be exposed to water stress caused by global climate change (Stern, 2007).

The poorest countries have in addition; can deal with the economic consequences of global climate change. Global climate change weakens States and reduces their ability to supply opportunities and services to help people recede vulnerable, particularly if those people already exist in marginalized areas. Indeed, the economy of the various developing countries is essentially supported agriculture and 1st product that are one in every of the foremost sectors directly touched by climate change and natural disasters. Further as their impoverishment, developing countries are in a disadvantageous and unsuitable situation thanks to their population growth, their giant urbanization and their geographical surroundings, that make them further vulnerable and fewer able to adapt to global climate change. It decreases gross domestic product growth, can increase the deficit and conjointly the external debt of states sometimes already weakened economically. Moreover, their low financial gain levels and their underdeveloped financial markets work impossible insurances and credits to cover them simply just in case of climatically shocks, that increases their vulnerability at individual and national levels. Global climate change is, then, a constraint to the action of the Millennium Development goals and to property development generally (Stern, 2007).

Many developing countries are experiencing an increase inside the frequency and costs of natural disasters that are determinable on the average at five-hitter of their value between 1997 and 2001 (IMF, 2003). In the Republic of India and South East Asia, the reduction in value to international climate change is determinable at between 9–13% by 2100 compared with a scenario without climate change. The cost of adaptation for these countries are going to be a minimum of between 5–10% of GDP and may weak the government budgets, all the more so since less than 1% of losses from natural disasters were insured in low-income countries for 1985 to 1999.

The frequency of climate events does not provide time to make or reconstitute their patrimony, keeping them in an exceedingly poor, there is a crowding-out impact as a results of the poorest are supposed to apportion their resources to influence the implications of global climate change instead of finance in human value like children's education or completely different productive investments. Immediate and sturdy reactions are then necessary for these specific countries to limit the extreme impact of global climate change on them. They suffer a double penalty as a results of, inside the present context, less-developed countries is fatherly treed in an exceedingly vicious circle: their impoverishment makes them additional vulnerable inside the face of global climate change and since of their impoverishment, climate change will have serious consequences on health, financial gain, and growth prospects and may trigger their impoverishment and vulnerability. In spite of this instance represented previously, international climate changes unfortunately thought-about as a protracted drawback and future impacts of climate change do not have priority.

Concerning now, Ikeme (2003) analyzes the low capability adaptation of SSA countries to deal with climate change effects. Indeed, low adaptive capability can increase vulnerability, social, and economic costs that have a bearing on human capital and conjointly the event levels of these areas that represent transmission channels for migration. For these countries, adaptation does not appear to be a pressing issue and is underestimated by these most vulnerable countries. Indeed, though adaptation is globally recognized as some way to preventing and handling the impacts of climate change, there is a relative in distinction and lean measures therefore on strengthen the aptitude of adaptation. Indeed, they are sometimes in an exceeding hard context with problems like impoverishment, institutional weakness, low levels of education associate degreed skills or associate degree inexistence of welfare systems; they are then supposed to act in an emergency simply just in case of climate effects (Washington et al., 2006). Moreover, developing countries, considerably in a federal agency, ponder the developed countries to be the foremost necessary rationalization for climate change, and wish to permit them to require the responsibility to manage them.

8.4   WHAT IS AGRIBUSINESS?

Agribusiness may be a broad proof of the notion that covers input suppliers, agro-processors, traders, exporters, and retailers. Agribusiness provides inputs to farmers and connects them to shoppers through the finance, handling, processing, storage, transportation, offer for sale and distribution of agro-industry merchandise, and can be rotten any into four main groups:

- Agricultural input manufacture for increasing agricultural productivity, like agricultural machinery, gear, and tools; fertilizers, insecticides; irrigation systems and connected equipment;
- Agro-industry: Food and beverages; tobacco merchandise, lacing, and animal skin products; textile, footwear, and garment; wood and wood merchandise; rubber merchandise; any as business products supported agricultural materials;
- Instrumentation for operation agricultural raw materials, likewise as machinery, tools, storage facilities, cooling technology and spare parts; and
- Varied services, financing, sale, and distribution firms, likewise as storage, transport, ICTs, packaging materials and modality for best promoting and distribution.

Agribusiness is thus a term used to mean farming and every one the alternative industries and services that represent the provision chain from the farm through operation, wholesaling, and promoting to the consumer (from farm to fork inside the case of food products).

Agro-industry includes all the post-harvest activities that square measure involved inside the transformation, preservation, and preparation of agricultural production for treated or final consumption of food and non-food merchandise (Wilkinson and Rocha, 2009). It consists of six main groups per the International customary Industrial Classification (ISIC), specifically food and beverages; tobacco products; paper and wood products; textiles, footwear, and apparel; animal product merchandise; and rubber merchandise. The term captures a numerous vary of primary and secondary post-harvest activities, ranging from basic village-level artifact preparation to fashionable process and involving wide differing levels of scale, quality, and labor, capital, and technology intensity. Food method industries tend to dominate this sector in developing countries, likewise Africa.

Rao (2006) groups food operation industries into three categories: (1) primary—those that involve the elemental operation of natural transition up, for example, cleaning, grading, and dehiscing; (2) secondary—those that embody simple or elementary modification of natural transition up, for example, chemical alteration of edible oils; and (3) tertiary—those that embody some kind of advanced modification to the natural end up like making it into edible merchandise like tomatoes into favorer, farm merchandise into cheese, etc.

The agri-food system encompasses the interlinked set of activities that run from seed to table, likewise as agricultural input production and distribution, farm-level production, raw product assembly, conversion, and merchandising. It encompasses the price chains for numerous agricultural and food merchandise and inputs and conjointly the linkages among them. The agri-food system is, in addition, a shorthand term for agriculture and connected agro-industries. Whereas most of the analysis refers expressly to it an element of this expanded agriculture that produces food, many of the conclusions apply equally well to those components of agriculture and agro-industry that end up non-food merchandise like fibers and biofuels.

Agro-processing is that the subset of processing that processes raw materials and intermediate merchandise derived from the agricultural sector. Agro-processing business thus suggests that reworking merchandise originating from agriculture, biology, and fisheries (FAO, 1997).

8.5   THE CHALLENGE: SPURRING PRODUCTIVITY GROWTH IN AGRO-FOOD SYSTEMS

Since 2000, there has been growing political and academic agreement that an agriculture-led approach to development, involving stronger productivity growth throughout the full agro-food system, offers the best likelihood for quick, widely-shared process and impoverishment reduction in SSA (Partnership to Cut Hunger and Poverty in Africa, 2002; Inter Academy Council, 2004; World Bank, 2007; UNECA and African Union (AU), 2007; Staatz and Dembélé, 2008).

Farming, small-scale agro-processing and promoting are notably necessary sources of women's incomes (World Bank et al., 2009). A body of empirical proof suggests that agricultural growth is that the key determinant of overall economical process and impoverishment reduction in most desert African countries (Christiansen and Demery, 2007; Byerlee et al., 2005; DFID, 2005; Dercon, 2009; Diao et al., 2003; Mwabu and Thorbecke, 2004; Wolgin, 2001). The growing political agreement relating to the importance of agriculture reflects the revived commitment of African governments and their development partners to supporting agrifood system development via initiatives just like the wonderful Class Africa Agriculture Development Program (CAADP), launched in 2003, and supported by African Heads of State, Government, and by most major development partners. CAADP sets daring goals for agricultural investment and growth. Member states of the AU have pledged to require a grip a minimum of 10% of fund resources to the agricultural sector, and conjointly the G-8, meeting in L'Aquila, Italian in 2009 revived donor commitments to CAADP (AU and NEPAD, 2004; G-8, 2009).

8.6   LOW PRODUCTIVITY IN SOUTHEAST MEDITERRANEAN SEA FARMING

In 2004, CAADP set a target for a mean annual rate of agricultural growth of 6%– the expansion rate that was judged necessary at the time if SSA countries were to fulfill the MDG1 of halving the impoverishment rate by 2015. Today, it's recognized that the overwhelming majority of states will not reach the MDG impoverishment reduction goal by 2015 (World Bank, 2010a), but the 6% agricultural rate remains a target.

Reaching this 6% annual rate poses a huge challenge. For sub-Saharan African countries to understand and sustain this target rate would need inflated productivity, not merely in farming but throughout the full agro food system. Out of 45 countries that info are out there, only one, Angola, throughout its post-war recovery half, achieved degree annual rate in agriculture value further between 2000 and 2008 of a minimum of 6% (World Bank, 2010b). Consistent with FAO estimates for Sub-Saharan Africa, agricultural production has grown up a lot of slowly than population over the past 45 years, with a succeeding decline in per capita food accessibility from domestic sources. (UNIDO et al., 2008) show that inflated agricultural production has been perform of further intensive land use—the growth of the world below crops or supporting livestock—arising from the relative abundance of cultivatable land and relatively low rents. Yields per area unit are essentially stagnant, notably for cereals, in distinction with substantial yield can increase in numerous regions. Cereal yields inflated alone 29% inside the 43 years between 1961–63 and 2003–05, compared to 177% in developing Asia and 100 and 44% in Latin America. The augmented farm production in countries that successfully exploited green revolution technologies was achieved by technological progress therefore total factor productivity (TFP) was the key offer of output growth, whereas in Africa the utilization of total inputs has been further necessary. Whether or not or not Africa can still increase agricultural production primarily through growth of land below cultivation is not as straightforward as sometimes urged. Whereas many reports (e.g., from the FAO) cite large areas of uncultivated cultivatable land, they seldom analyses the economic and environmental costs of conveyance such land below production. Economic costs someone the investment costs of infrastructure any as a result of the costs of human and illness management necessary to open these areas to farming.

Potential infrastructure costs are staggering: the proportion of land below irrigation in SSA is presently less than 1/4 of the Republic of India in 1961, at the dawn of its green revolution. Increasing the share of irrigated land in the federal agency to Indian levels in 1960 would value roughly $114 billion. Similarly, SSA's road density, at 201 km/1000 km^2, maybe a smaller quantity than a third of that of the country in 1950 (703 km/1000 km^2). Even Rwanda, the continent's most densely peopled country, does not have the road density of the Republic of India in 1950. Today's gap is even wider: India's road density is 32-fold that of Ethiopia and 255 times that of the Sudan (Staatz and Dembélé, 2008). Moreover, agricultural house growth in SSA sometimes involves deforestation (with implications for world climate change) and loss of important life habitats. Thus, whereas some countries, like Zambia, may need a scope for property agricultural growth through agricultural growth, the very important question facing the continent is that the relative cost of house growth versus augmentative production on existing land.

The low productivity of agriculture is partially performed of the low-level of use of industrialized inputs. However, 4% of desert Africa's cultivatable land is irrigated (compared with nearly 39% in South Asia and 11% for Latin America and also the Caribbean); leading to each lower and a lot of unstable yields for many major staple crops. Similarly, the intensity of chemical and agricultural machinery use is one-eighth to a minimum of one-tenth of that in South Asia. Intensity of manufactured input use varies wide across the continent; West Africa's use of these inputs may be a number of third of that of states inside the South African Development Community (SADC) region and between a 20% and 25% of that in countries within the Common Market for Eastern and Southern Africa (COMESA) (UNECA and AU, 2009). In part, this reflects lower population densities and higher inherent soil fertility in some components of

West Africa any as a result of the larger incidence of large-scale farming in the Republic of South Africa and Namibia. Over the long run, increasing the utilization of inputs like chemical are visiting be very important to increasing farm-level productivity, incomes. A 2009 World Bank study on the fight of business agriculture in Africa compared the on-farm per-unit production prices for several agricultural merchandise created inside the Guinea-Savannah regions of Africa with production costs for constant merchandise in Brazil and Asian countries. The study showed that whereas African farm-level costs were like those in Brazil and Asian countries, this 'competitiveness' was supported soil mining (the depletion of soil nutrient reserves, leading to soil degradation) and extremely low returns to labor, reflective few numerous employment opportunities for workers—hardly a model for impoverishment reduction (World Bank, 2009). Whereas labor productivity in agribusiness varies significantly amongst entirely completely different the African countries, not alone is productivity low by international standards it's stagnated over time. Productivity levels in African business enterprise are low part as a result of educational levels fall well wanting the standard required to achieve technical efficiency in agriculture and manufacturing. In rural areas in SSA, North Africa, South Asia and conjointly the Middle East, adult males have regarding four years of education and females even less (1.5 to four years), whereas in Central Asia and Europe education levels are upper (World Bank, 2007). Health and accomplishment standards are equally poor.

8.7 VALUE ADDITION IN THE CONTEXT OF STRUCTURAL TRANSFORMATION

Value addition inside the agro-food system is commonly confused with the method that changes the form of the production. Value is commonly further to merchandise whereas not ever dynamical their physical kind and method (in the sense of fixing the form of the merchandise) do not basically add value to the product. Value addition can involve method inside the sense that the merchandise undergoes some technique (which can merely involve cleanup, grading, or labeling), once that an emptor is willing to pay a price for the merchandise that overcompensates for the worth of the inputs used within the technique. Sorting a heterogeneous mixture of mangoes into high-quality fruit targeted to the modern fruit export market and lower-quality fruit targeted to juice production for native consumption permits a firm to separate markets and apply price discrimination by charging higher value in export markets for high-quality modern mangoes, thereby increasing its earnings. In an exceedingly very economic system, this extra price is typically manifested by the processor earning a profit (Haggblade et al., 2011).

In a disparity, wherever method uses resources that are price over the additional amount customers are willing to obtain this processed product relative to the raw product, value subtraction rather than value addition happens. A classic example in West Africa was the development of fresh abattoirs in several Sahel Ian countries throughout the 1960s and 1970s, with the aim of exportation cool meat to coastal countries. The motivation was to substitute the export of fresh meat from the Sahel for the export of live animals (cattle, sheep, and goats) to the coast, thereby getting the value further of operation at the livestock exportation countries. In practice, the bulk of these efforts failing, partly as a result of cold meat transport from the West Africa to the Sahel was further costly and unreliable than the transport of live animals (Staatz, 2011).

Another necessary not withstanding underappreciated issue was the plenty of upper value that may be attained on the coast for the organs and completely different by-products of the slaughtered animal (hides, hoofs, horns, etc.), the supposed 'fifth quarter.' as a results of coastal residents consume as food or retread a lot of these merchandise than do Sahel Ian, they are willing to pay a so much higher value for them, and since many of these by-products are pure fable, they are exhausting to ship from West Africa to the Sahel whereas not intensive processing—unless they are shipped as an element of a living animal. As a consequence, traders involved in the export of live animals could afford to pay further for export-quality animals than could firms involved inside the meat export business. Native slaughter for export became a money-losing proposition—a value subtraction activity instead of value addition (Makinen et al., 1981).

8.8   VALUE ADDITION, AGRO-INDUSTRY, AND THE PROCESS OF STRUCTURAL TRANSFORMATION

Nearly each economy, where living standards have up well has undergone structural transformation, whereby the proportion of the complete population engaged in farming falls as can farming's relative contribution to worth, inside the tip of the day, structural transformation involving a net resource transfer from farming to completely different sectors of the economy is significant for impoverishment reduction in SSA, as somewhere between one and two-thirds of small farmers (depending on the country) appear to lack the resources to farm their route out of poverty and may, therefore, would love eventually to maneuver to further remunerative employment outside farming (Staatz and Dembélé, 2008). Structural transformation involves a reorientation of the economy off from subsistence-oriented, household-level production associate degreed household-based agro-industry towards an integrated economy supported larger specialization; exchange and capturing of economies of scale (Reynolds, 1985). Many functions long ago conducted on the farm, like input production and output method, are shifted to off-farm elements of the economy. Farmers depend further on external sources of power (e.g., diesel-powered pumps) and fewer on producing that power themselves through human or animal power. Thus, resources shift among the agro-food system, any as between the food system and conjointly there minder of the economy, therefore, off-farm players inside the food system—agribusiness and food retailing—grow relative to farm-level production in terms each of import further and employment.

One implication of this technique is that driving down the real value of food to consumers—critical to impoverishment reduction as a result of low-income Africans pay a high proportion of their incomes on food—demands augmented attention to fostering technical and institutional changes in every off-farm agro-food activities and farm production. Increasing farm productivity may be a necessary, but not an adequate condition to cut back the necessary price of food and to substantiate that African agribusinesses are competitive internationally. A second implication is that reduced human activity prices are a necessity for structural transformation. High human activity prices (difficulties relevant judgment and contract group action and demands for bribes at borders) can choke off structural transformation by making it too expensive for people to depend upon the specialization and exchange necessary to need the advantage of recent agro-food technologies. Structural transformation, in addition, involves the inflated integration of actors inside the agro-food system into broader, sometimes the world, knowledge systems. Invariably such knowledge is embodied in new technologies, management practices, institutions, and adept networks that develop and exchange such knowledge. As economies retread, the process depends more and more on embodied knowledge and knowledge transfer. Structural modification among agriculture is dominated by the shift from heterogeneous, but subsistence-oriented farms, towards further specialized market-driven production. Consequently, transformation involves inflated integration of farming and agro-industry, any as of the full agro-food system with completely different sectors of domestic and world economies. Inside the first stages of agricultural-led growth, agricultural production and exports are typically dominated by bulk agricultural commodities, that resource endowment (and transport infrastructure) are key determinants of comparative advantage. As countries move further into the production of higher price agro-food merchandise, competitive advantage is more and more determined by investments in human capital, analysis, development, and logistics (Abbott and Brehdahl, 1993).

8.9   MAKING AGRIBUSINESS A PRIORITY

Agricultural development has the potential to bolster food security, turn out opportunities, and raise incomes for the world's poor, many of whom board rural farm communities. Sustainability in agriculture may be a business driver for IFC's purchasers and partners. IFC is achieving this through:

- Enhancing sustainability;
- Increasing production employing fewer resources and reducing the impact on the environment;
- Improving the use of energy and water effectiveness;
- Increasing productivity;
- Building skills and providing higher seeds and inputs;
- Assisting farmers adopt widespread agricultural practices;
- Enabling access to finance;
- Facilitating development and availableness of applicable money merchandise like weather-based insurance to include risks;
- Building capability of economic institutions to cater to the requirements of little farmers;
- Facilitating access to markets;

- Assisting farmers to meet the number and quality needs of larger markets;
- Strengthening storage and reposting infrastructure to cut back field-to-market losses.

8.10   SOUTHEAST MEDITERRANEAN SEA: WHY AGRIBUSINESS?

Manufacturing has not played about a dynamic role inside the economic development of Southeast Mediterranean up to currently. There are pressing issues that call for a reorientation to support agribusiness and agro-industrial development: specifically, impoverishment reduction and conjointly the action of the MDGs; and ensuring simply patterns of growth, which can address the concentration of employment and livelihoods inside the agricultural sector. An agribusiness development path involving larger productivity growth throughout the full business worth chain—covering farms, firms, and distributors—represents a solid foundation for quick, inclusive process and impoverishment reduction (Yumkella et al., 2011).

8.11   AGRIBUSINESS, ECONOMIC GROWTH AND POVERTY REDUCTION

Alongside its role in stimulating economic development, agribusiness, and agro-industrial, development has the potential to contribute well to impoverishment reduction and improved social outcomes and an accord is rising that agro-industries square measure a decisive a part of socially-inclusive, competitive development strategies (Wilkinson and Rocha, 2008). Proof of the link between growth and impoverishment reduction varies per country. Spectacular economic and industrial growth in China lifted 475 million people out of impoverishment between 1990 and 2005, though large pockets of impoverishment still exist in growth-oriented areas and rural communities to structural rigidities. In SSA, despite durable growth in recent years, the number of people living on however $1.25 on each day inflated by 93 million throughout the same period (Montalvo and Raval-lion, 2010; World Bank, 2009).

The implementation of the MDGs in SSA has been forced by two factors: first, most countries haven't met the required worth rate to reach the MDG1 target. Secondly, labor absorption and employment intensity are low to the level of growth in some capital-intensive extractive sectors. Agribusiness directly contributes to the action of three key MDG's, specifically reducing impoverishment and hunger (MDG1), empowering girls (MDG3), and developing international partnerships for development (MDG8). Strong synergies exist between business, agricultural performance and impoverishment reduction for Africa (World Bank, 2007); economical business would possibly stimulate agricultural growth and sturdy linkages between agribusiness smallholders can in the reduction of rural impoverishment. Attention on value addition in agribusiness is, therefore, central to existing strategies for economic diversification, structural transformation, and technological upgrading of African economies. Such attention can initiate faster progress towards prosperity, by poignant the bulk of the continent's economic activities and by harnessing important linkages between the foremost important economic sectors. This 'people-oriented' strategy will improve welfare and living standards of the overwhelming majority of Africans, every as producers and shoppers, and from the angle of employment, it ought to be tough to lift people out of impoverishment through direct employment generation in manufacturing alone, even with high employment snap and productivity growth, until the low initial manufacturing base can increase in importance. The indirect impact of manufacturing growth on employment, however, is probably going to be quite very important, notably inside the context of backward and forward linkages to agriculture and services through business development. In separate agencies, such linkages are doubtless to possess nice potential, despite being presently rather weak to a restricted proportion of agricultural raw materials being processed and exported and since very important industry-related services won't, however, have developed, gain, and food security. On the demand aspect, food expenditure sometimes represents the foremost necessary single item of family expenditure, rising to over half total expenditure for poor households in some countries, and therefore the efficiency of post-harvest operations may be a significant determinant of the costs paid by the urban and rural poor for food, and thus a significant consider family food security (Jaffee et al., 2003).

Agro-industrial development can contribute to boost health and food security for the poor by increasing the general handiness, variety, and nutritional value of food products, and enabling food to be kept as a reserve against times of shortage, ensuring that adequate food is accessible there and that essential nutrient are consumed throughout the year. On the availability aspect, agro-industrial development options and on the spot impact on the livelihoods of the poor each through augmented employment in agro-industrial activities, and thru augmented demand for primary agricultural end up. Though variable significantly by subsector and region, agro-industry, considerably in its initial stages of development, is relatively labor-intensive, providing a ramification of opportunities for self and wage employment, inside the case of Siam, for example. Watanabe et al. (2009) counsel that, for the period 1988–2000 not alone was the amount of workers per worth extra for agro-processing at or on top of the mean for producing overall. The amount of poor workers per worth extra within the agro-processing business was considerably larger. The figures for food merchandise (and the smaller wood and wood merchandise sector) were over double those of the standard of the manufacturing business, implying that the agro-processing business, considerably the food business, tends to rent a much bigger varies of the poor than completely different manufacturing industries.

Agro-industrial activity in Africa is in addition typically distinguished by a high proportion of female employment, ranging from 50% to as high as 90%, these figures exclude metals and energy. It should be noted that, among some market sectors a minimum of, strong gender segmentation in production and process tends to consign girls to a lot of vulnerable kinds of work (casual, temporary, and seasonal), lower-paid and extra labor-intensive preparation and/or processing (UNIDO et al., 2008; Sir Geoffrey Wilkinson and Rocha, 2008). For example, the 'non-traditional export sector' (vegetables, fruit, and fish products), that's presently the foremost dynamic in terms of exports from SSA. Similarly, the small-scale food process and line of labor operations gift throughout plenty of the continent are sometimes operated predominantly by ladies; a study of small-scale urban agro-processing and business enterprises in Cameroon found that over 80% were managed by girls, with men being gift nearly solely within the mechanical milling/grinding and meat preparation activities (Ferré et al., 1999).

Indeed, Charmes notes that the gender bias apparent in many agro-processing activities would possibly contribute to the general approximation of every agro-industrial activity and female employment in national accounting, noting that a very high share of these activities are below taken as secondary activities and typically hidden behind subsistence agriculture

(Charmes, 2000). Aboard job creation, agro-industrial enterprises some-times provide crucial inputs and services to the farm sector for those with no access to such inputs, inducement productivity and merchandise quality enhancements and stimulating market elicited innovation through chains and networks, facilitating linkages and allowing domestic and export markets to become more reciprocally supportive (FAO, 2007). Agro-industry is additionally amongst the foremost accessible of business activities–frequently undertaken at small-scale, with low initial value and technological barriers to entry. SMEs keep key actors inside the largely informal transaction and method networks, that dominate food acquisi-tion in an exceedingly ton of (newly) urban Africa, and have established remarkably adaptive and resilient inside the face of a ramification of economic, institutional, and infrastructural challenges (Muchnik, 2003; Sautier et al., 2006).

8.12   REGIONAL IMPACTS OF AND VULNERABILITIES TO CLIMATE CHANGE

Africa is already a continent fraught from climate stresses and is incredibly at risk of the impacts of climate change. Several areas in Africa are recog-nized as having climates that are among the foremost variable inside the globe on seasonal and decadal time scales. Floods and droughts will occur inside the identical place among months of each completely different. These events can cause famine and a wide unfold disruption of socio-economic well-being. For example, estimates rumored indicate that one-third of African people already board drought-prone space and 220 million are exposed to drought annually. Several factors contribute and compound the impacts of current climate variability in Africa and can have nega-tive effects on the continent's ability to deal with global climate change. These embody impoverishment, illiteracy, and lack of skills, weak institu-tions, restricted infrastructure, lack of technology and data, low levels of primary education, and health care, poor access to resources, low manage-ment capabilities, and armed conflicts. The event of resources likewise as forests can increase in population, action, and land degradation produces any threats (UNDP, 2006). Within the Sahara and Sahel, mud, and sand storms have negative impacts on agriculture, infrastructure, and health. As a result of warming, the climate in Africa is foreseen to become further variable, and extreme weather events are expected to be further frequent and severe, with increasing risk to health and life. This includes increasing risk of drought and flooding in new areas (Few et al., 2004; Christensen et al., 2007) and inundation to sea-level rise inside the continent's coastal areas (Nicholls, 2004; McMichael et al., 2004). As an effect of worldwide (climate change global climate change temperature change) in Africa on key sectors and offers an indication of the adaptive capability of this continent to climate change.

Africa can face increasing water lack and stress with a succeeding potential increase of water conflicts as most of the 50 river basins in Africa are Tran's boundary (Ashton, 2002; Wit and Jacek, 2006). Agricultural production depends chiefly on downfall rain for irrigation and may be severely compromised in many African countries, notably for subsistence farmers in SSA. Beneath international global climate change, plenty of agricultural lands are visiting, be lost, with shorter growing seasons and lower yields. National communications report that international global climate change will cause a general decline in most of the subsistence crops, e.g., sorghum in Sudan, Ethiopia, Eritrea, and Zambia; maize in Ghana; Millet in Sudan; and groundnuts in the Gambia. Of the complete further people in peril of hunger due to international global climate change, although already associate degree outsize proportion, Africa could account for the majority by the 2080s (Fischer et al., 2002). Africa is at risk of sort of climate-sensitive diseases together with malaria, tuberculosis, and diarrhea (Guernier et al., 2004). Below international global climate change, rising temperatures are dynamic the geographical distribution of illness vectors that are migrating to new areas and better altitudes, as associate degree example, migration of the mosquito to higher altitudes can expose giant numbers of antecedently unexposed folks to infection within the densely inhabited geographic region highlands (Boko et al., 2007). Future climate variability also will act with different stresses and vulnerabilities like HIV/AIDS (which is already reducing anticipation in many African countries) and conflict and war (Harrus and Baneth, 2005), leading to augmented susceptibility and risk to infectious diseases (e.g., epidemic cholera and diarrhea) and malnutrition for adults and kids (WHO, 2004).

Climate change is another stress to already vulnerable habitats, ecosystems, and species in Africa, and is perhaps running trigger species migration and cause environment reduction; up to 50% of Africa's total multifariousness is in peril due to reduced environment and various human-induced pressures (Boko et al., 2007). The latter embraces land-use conversion due to agricultural growth and succeeding destruction of habitat; pollution; poaching; civil wars; high rates of land-use change; growth and the introduction of exotic species. As an example, it is perhaps visiting considerably decline, at the western lowland between 2002 and 2032. Future water level rise has the potential to cause vast impacts on the African coastlines likewise because of the already degraded coral reefs on the Japanese coast. National communications indicate that the coastal infrastructure in 30%of Africa's coastal countries, together with the Gulf of Guinea, Senegal, Gambia, Egypt, and on the East-Southern African coast, is in danger of partial or complete inundation to accelerated water level rise. In Tanzania, a water level rise of 50 cm would inundate over 2,000 km² of land, estimate accounting around USD 51 million (UNEP, 2002). Future water level rise, in addition, threatens lagoons and mangrove forests of every Nipponese and western Africa, and is perhaps going to impact urban centers and ports, like port Maputo, and Dar El-Salaam.

8.13 CONCLUSION

This chapter provides a quick summary of background of promoting agribusiness development in Southeast Mediterranean. Where, the adoption of the 2030 Agenda marked the beginning of a new era with a strong commitment from the international community to promote a wide range of transformative and universal changes to achieve Sustainable Development, with local and regional specificities. The Sustainable Development Goals (SDGs) strive to provide inclusion and empowerment for all. The operationalization of this inclusive approach to growth and development relies on integrating the economic, social and environmental dimensions of development. Agriculture and food security play a key role in this regard. Indeed, they are at the heart of the 2030 Agenda. Where the world, including the Mediterranean region, is faced with a number of challenges, such as inequalities, significant flows of distress migration and limited access to and poor management of natural resources, including water, land, and biodiversity.

In terms of food, the world produces enough today to feed the planet, but one third, representing 1.3 billion tons per year, is either wasted or lost in the supply chain, from initial agricultural production all the way to final household consumption. Furthermore, continued increase in the use of natural resources such as water, land, forestry, biodiversity and fisheries, without paying sufficient attention to their depletion or environmental impacts, can lead to ecological crises and security threats. In the Mediterranean region, for example, wasting a precious resource like water may intensify such threats. Additionally, the waste of human resources hampers development efforts. This happens, for example, in the form of unemployment, lack of access to education especially for girls, "brain drain" from developing countries, disappearance of local knowledge such as family farming practices and products, duplication of ideas without coordination and lack of synergies among relevant actors. This represents an important step towards building consensus on innovations and inclusive policies needed to respond to the challenges faced by the Southeast Mediterranean region, particularly in terms of the achieving food security related to natural resources, food and knowledge. I believe that this chapter contributes to fostering synergies in thematic areas of mutual interest. We hope that this joint piece of work will act as a catalyst for action towards achieving food security and sustainable development in the region, in collaboration with policymakers and all the other actors of the Euro-Mediterranean multilateral cooperation.

KEYWORDS

- **catastrophe**
- **crop yields**
- **global climate change**
- **global data**
- **International poverty line**
- **World Bank**

REFERENCES

Abbott, P. C., & Brehdahl, M. E., (1993). Competitiveness: Definitions, useful concepts, and issues. In: Brehdahl, M. E., Abbott, P. C., & Reed, M., (eds.), *Competitiveness in International Food Markets.*

African Union and NEPAD, (2004). *Implementing the Comprehensive Africa Agriculture Development Program and Restoring Food Security in Africa*. The Roadmap, Midland, South Africa: NEPAD Secretariat.

Ashton, P. J., (2002). Avoiding conflicts over Africa's water resources. *Ambo, 31*(3), 236–242.

Blas, J., & Meyer, G., (2010). *Agribusiness: All You Can Eat Financial Times*.

Boko, M., Niang, I., Nyong, A., Vogel, C., Githeko, A., Medany, M., Osman-Elasha, B., Tabo, R., & Yanda, P., (2007). Africa. In: Parry, M. L., Canziani, O. F., Palutikof, J. P., Van der Linden, P. J., & Hanson, C. E., (eds.), *Climate Change 2007: Impacts, Adaptation, and Vulnerability: Contribution of Working Group II to the Fourth Assessment Report of the Intergovernmental Panel on Climate Change* (pp. 433–467). Cambridge University Press. Cambridge UK.

Byerlee, D., Xinshen, D., & Jackson, C., (2005). *Agriculture, Rural Development, and Pro-Poor Growth: Country Experiences in the Post-Reform Era*. Agriculture and rural development discussion paper 21, Washington, D.C.: World Bank.

Charmes, J., (2000). *African Women in Food Processing: A Major, But Still Underestimated Sector of Their Contribution to the National Economy*. Paper prepared for the international development research Centre (IDRC), Ottawa: IDRC.

Christensen, J. H., Hewitson, B., Busuioc, A., Chen, A., Gao, X., Held, I., et al., (2007). Regional climate projections. In: Solomon, S., Qin, D., Manning, M., Chen, Z., Marquis, M., Averyt, K. B., Tignor, M., & Miller, H. L., (eds.), *Climate Change 2007: The Physical Science Basis: Contribution of Working Group I to the Fourth Assessment Report of the Intergovernmental Panel on Climate Change*. Cambridge University Press, Cambridge, United Kingdom and New York, NY, USA.

Christiansen, L., & Demery, L., (2007). *Down to Earth: Agriculture and Poverty Reduction in Africa*. Washington D.C.: The World Bank.

De Wit, M., & Jacek, S., (2006). *Changes in Surface Water Supply Across Africa with Predicted Climate Change*. Africa Earth Observatory Network (AEON), University of Cape Town. Rondebosch 7701, South Africa Science Express Report.

Dercon, S., (2009). Rural poverty: Old challenges in new contexts. *The World Bank Research Observer, 24*(1), 1–28.

DFID, (2005). Department for international development. *Growth and Poverty Reduction: The Role of Agriculture*. DFID Policy Paper, London: DFID.

Diao, X., Dorosh, P., Rahman, S. M., Meijer, S., Rosegrant, M., Yanoma, Y., & Li, W., (2003). *Market Opportunities for African Agriculture: An Examination of Demand-Side Constraints on Agricultural Growth*. Development Strategy and Governance Division Discussion, Paper Series 1. Washington, D.C.: IFPRI.

FAO, (1997). *Food and Agriculture Organization of the United Nations*. Available at: http://www.fao.org/docrep/w5800e/w5800e00.htm (accessed on 4 January 2020).

FAO, (2007). Food and Agriculture Organization of the United Nations. *Challenges of Agribusiness and Agro-Industry Development* (pp. 1–13). Boulder: Westview Press. Available at: ftp://ftp.fao.org/docrep/fao/meeting/011/j9176e.pdf (accessed on 4 January 2020).

Ferré, T., Doassem, J., & Kameni, A. (1999). Dynamics of agricultural product processing activities in Garoua, North Cameroon, Garoua, Cameroon: Institute for Agricultural

Research for Development (IRAD)/Pole of Research Applied to Development of Savannas of Central Africa (PRASAC).

Few, R., Ahern, M., Matthies, F., & Kovats, S., (2004). *Floods, Health and Climate Change: A Strategic Review.* Working Paper No. 63. Tyndall Center for Climate Change Research.

Fischer, G., Mahendra, S., & Harrij, V. V., (2002). Climate change and agricultural variability: A special report. In: *On Climate Change and Agricultural Vulnerability: Contribution to the World Summit on Sustainable Development.* Johannesburg (Global, agriculture).

G-8, (2009). *L'Aquila Joint Statement on Global Food Security: L'Aquila Food Security Initiative (AFSI).* Available at: http://www.g8italia2009.it/static/G8_Allegato/LAquila_Joint_Statement_on_Global_Food_Security%5b1%5d,0.pdf (accessed on 4 January 2020).

Guernier, V., Hochberg, M. E., & Guegan, J. F., (2004). Ecology drives the worldwide distribution of human diseases. *PLOS Biology, Oxford, 2*(6), 740–746.

Hamada, Y. M. (2018). Special pictures album, Took by Youssef M. Hamada, Egypt, 2018.

Harrus, S., & Baneth, G., (2005). Drivers for the emergence and re-emergence of vector-borne protozoal and rickettsial organisms. *International Journal for Parasitology, 35*, 1309–1318.

Ikeme, J., (2003). Climate change adaptation deficiencies in developing countries: The case of sub-Saharan Africa. *Mitigation and Adaptation Strategies for Global Change, 8*(1), 29–52.

IMF, (2003). *Fund Assistance for Countries Facing Exogenous Shocks.* Prepared by the Policy Development and Review Department (In consultation with the area, finance and fiscal affairs departments) International Monetary Fund, Technical report, Washington, DC, IMF.

Inter Academy Council, (2004). *Realizing the Promise and Potential of African Agriculture.* Amsterdam: Inter Academy Council.

Jaffee, S., Kopicki, R., Labaste, P., & Christie, I., (2003). *Modernizing Africa's Agro-Food Systems, Analytical Framework and Implications for Operations.* Africa Region Working Paper Series No. 44, Washington D.C.: The World Bank.

John, S., (2011). *Agribusiness for Africa's Prosperity.* UNIDO ID/440, Layout by Smith + Bell Design (UK), Printed in Austria.

Makinen, W., Herman, L., & Staatz, J., (1981). *A Model of Meat versus Live-Animal Exports from Upper Volta.* CRED discussion paper no. 80, Ann Arbor: University of Michigan Center for Research on Economic Development.

McMichael, A. J., Campbell-Lendrum, D., Kovats, R. S., Edwards, S., Wilkinson, P., Edmonds, N., et al., (2004). In: Ezzati, M., Lopez, A. D., Rodgers, A., & Murray, C. J. L., (eds.), *Climate Change in Comparative Quantification of Health Risks: Global and Regional Burden of Disease Due to Selected Major Risk Factors* (Ch. 2, pp. 1543–1649). World Health Organization, Geneva.

Montalvo, J., & Ravallion, M., (2010). The pattern of growth and poverty reduction in China. *Journal of Comparative Economics, 38*(1), 2–16.

Muchnik, José (2003). *Food, Know-How and Agro-Food Innovations in West Africa,* Collection of reports from the ALISA project. European Union DG XII, Brussels, 2003.

Mwabu, G., & Thorbecke, E., (2004). Rural development, growth, and poverty in Africa. *Journal of African Economies, 13*, 16–65.

Myers, N., (2005). *Environmental Refugees, an Emergent Security Issue* (pp. 23–27). Paper presented to the 13th Economic Forum, Prague, Czech Republic.

Partnership to Cut Hunger and Poverty in Africa, (2002). *Now is the Time: A Plan to Cut Hunger and Poverty in Africa*. Washington D.C.: Partnership to Cut Hunger and Poverty in Africa.

Rao, K. L., (2006). Agro-industrial parks experience from India. In: *Agricultural and Food Engineering Working Document* (p. 3). Rome: FAO. Available at: http://www.fao.org/ag/ags/publications/docs/AGST_WorkingDocuments/J7714_e.pdf (accessed on 4 January 2020).

Rehdanz, K., & Maddison, D., (2005). Climate and happiness. *Ecological Economics, 52*(1), 111–125.

Reynolds, L., (1985). *Economic Growth in the Third World, 1850–1980*. New Haven: Yale University Press.

Robert, J. N., (2004). Coastal flooding and wetland loss in the 21st century: Changes under the SRES climate and socio-economic scenarios. *Global Environmental Change, 14*(2004), 69–86.

Saleemul, H., (2009). *The Anatomy of a Silent Crisis*. Global humanitarian forum; chair, steering group, human impact report: Climate change.

Sautier, D., Vermeulen, H., Fok, M., & Biénabe, E., (2006). *Case Studies of Agri-Processing and Contract Agriculture in Africa*. RIMISP, Latin American Center for Rural Development, Santiago: RIMISP.

Staatz, J. M., & Dembélé, N. N., (2008). *Agriculture for Development in Sub-Saharan Africa*. Background paper for the World Development Report. Washington D.C.: The World Bank.

Stern, N., (2007). *The Economics of Climate Change:* The stern review, Cambridge University Press.

Stern, N., Peters, S., Bakhshi, V., Bowen, A., Cameron, C., Catovsky, S., et al., (2006). *Stern Review: The Economics of Climate Change*. HM Treasury, Chapter 6, London. http://www.hm-treasury.gov.uk/stern_review_report.htm (accessed on 4 January 2020).

Steven, H., John, S., Duncan, B., Boubacar, D., Ferdinand, M., Isaac, J. M., Lulama, N. T., David, T., Ricker-Gilbert, J., Jayne, T. S., & Chirwa, E. V., (2011). Subsidies and crowding out: A double-hurdle model of fertilizer demand in Malawi. *American Journal of Agricultural Economics, 93*(1).

UNDP, (2006). Human *Development Report, Beyond Scarcity: Power, Poverty, and the Global Water Crisis*. United Nations Development Program. http://hdr.undp.org/hdr2006/report.cfm (accessed on 4 January 2020).

UNECA and African Union, (2007). *United Nations Economic Commission for Africa Economic Report on Africa 2007*. Accelerating Africa's development through diversification. Addis Ababa: UNECA.

UNECA and African Union, (2009). *United Nations Economic Commission for Africa (UNECA) and African Union*, Economic Report on Africa 2009: Developing African Agriculture through regional value chains. Addis Ababa: UNECA.

UNEP, (2002). *Africa Environment Outlook: Past, Present and Future Perspectives*. United Nations Environment Program. http://www.grida.no/aeo/ (accessed on 4 January 2020).

UNIDO, (2009). *United Nations Industrial Development Organization, Industrial Development Report 2009.* Breaking in and moving up: New industrial challenges for the bottom billion and the middle-income countries, Vienna: UNIDO.

UNIDO, FAO, IFAD, (2008). *The Importance of Agro-Industry for Socio-Economic Development And Poverty Reduction.* New York: UN Commission on sustainable development.

Washington, R., Harrison, M., Conway, D., Black, E., Challinor, A., Grimes, D., Jones, R., Morse, A., Kay, G., & Todd, M., (2006). African climate change: Taking the shorter route. *Bulletin of the American Meteorological Society, 87*(10), 1355–1366.

Watanabe, M., Jinji, N., & Kurihara, M., (2009). Is the development of the agro-processing industry pro-poor? The case of Thailand. *Journal of Asian Economics, 20*(4), 443–455.

WFP, (2009). *Who Are the hungry?* World Food Program, http://www.wfp.org/hunger/who-are (accessed on 4 January 2020).

WHO, (2004). *The World Health Report, 2003: Shapes the Future?* World Health Organization, Geneva. http://www.who.int/whr/2003/en/index.html (accessed on 4 January 2020).

Wilkinson, J., & Rocha, R., (2008). *Agro-Industries, Trends, Patterns, and Developmental Impacts.* Paper prepared for global agro-industries forum (GAIF), New Delhi.

Wilkinson, J., & Rocha, R., (2009). Agro-industry trends, patterns and development impacts. In: Da Silva, C., Baker, D., Shepherd, A. W., Jenane, C., & Miranda-da-Cruz, S., (eds.), *Agro-Industries for Development* (pp. 46–91), Wallingford, UK: CABI for FAO and UNIDO.

Wolgin, J. M., (2001). *A Strategy for Cutting Hunger in Africa.* Report commissioned by the partnership to cut hunger in Africa. Washington D.C.: Partnership to cut hunger and poverty in Africa.

World Bank, (2007). *World Development Report 2008: Agriculture for Development,* Washington, D.C.: World Bank.

World Bank, (2009). *World Development Indicators 2009.* Washington D.C.: The World Bank.

World Bank, (2010a). *Global Economic Prospects 2010: Crisis, Finance and Growth.* Washington D.C.: The World Bank.

World Bank, (2010b). *World Development Indicators 2010.* Washington D.C.: The World Bank.

World Bank, FAO, (2009). *Awakening Africa's Sleeping Giant: Prospects for Commercial Agriculture in the Guinea Savannah Zone and Beyond.* Washington D.C. and Rome: The World Bank and FAO.

World Bank, FAO, IFAD, (2009). *Gender in Agriculture Sourcebook.* Washington D.C.: The World Bank.

Yumkella, K. K., Kormawa, P. M., Roepstorff, T. M., & Hawkins, A. M., (2011). *Agribusiness for Africa's Prosperity.* UNIDO ID/440, Layout by Smith + Bell Design (UK), Printed in Austria.

CHAPTER 9

Potential Alleviate Poverty in Rural Areas for Climate Change Adaptation

ABSTRACT

Climate change poses an additional, serious challenge to increasing agricultural productivity in the Southeast Mediterranean Sea. Although models dissent in their plans of the rule and choice of effects, they typically elect three points (World Bank, 2007; Rocket scientist, 2007; UNIDO et al., 2008):

• The variability in weather events is maybe visiting rise, signifying higher risks for the agriculture sector and others. As an example, exaggerated risks of scarcities and floods as a result of higher temperatures are probable to steer to larger crop-yield losses.
• As a truth, the Southeast Mediterranean Sea is going to be more durable hit than completely different areas of the world by higher temperatures and reduced rain, partly for some crops grown in the Southeast Mediterranean Sea are already produced at the bounds of their heat tolerance.
• The impacts aslant Southeast Mediterranean Sea will not be even. As an example, the Sahel and regions of Southern Africa are presently becoming drying than among the past, though rain stages are attainable to rise in other areas, like parts of East Africa.

General, these world climate change effects are probable to rise the import dependency of the numerous Southeast Mediterranean Sea countries for some of their staple foods. They are put together probable to put enlarged compression on agricultural examination systems to develop diversities that are extra heat and drought-stress tolerant. Efforts to mitigate world climate change might, however, offer some new opportunities to Southeast Mediterranean Sea cultivation. If the institutional implementation arrangements could also be discovered to link Southeast Mediterranean Sea farmers to world carbon markets, there is an attainable for carbon sequestration among little farmers to become a really necessary new 'cash crop' in Southeast Mediterranean Sea, and better dominant of agricultural by-products and manures can end in larger native production of biogas to fuel farm and agro-processing operations (World Bank, 2007). During this chapter, new insights on potential alleviate impoverishment in rural areas for global climate change adaptation is going to be doing. This chapter goes over the matter of potentially alleviate impoverishment in rural areas for global climate change adaptation.

9.1 BACKGROUND

There is a tremendous heterogeneousness in agriculture across the Southeast Mediterranean Sea. What's additional, the figures for cereal yields might amplify the dearth of productivity growth in Southeast Mediterranean Sea agriculture as substantial diversification has taken place in some countries aloof from basic cereals into higher-value products and completely different staples, like cassava that have had recent major productivity-enhancing breakthroughs. Figure 9.1 shows the irrigation development projects in Upper Egypt.

To some, water-related world climate change might sound like an abstract development, but to Southeast Mediterranean Sea its impact is every real and immediate. On the border of the Sahara, where water has invariably been a precious and restricted resource, world climate change is accentuating the matter, wetting even scarcer as a result of decreasing rain and increasing droughts. Water deficiency not only threatens food production but put together has undercut the government's progress in increasing access to safe potable and up to sanitation. On the average, a heavy drought has occurred every 11 years for the past 100 years. However, over the past 30 years, drought frequency, intensity, and continuation have enlarged (WHO, 2006; Esper et al., 2007). Overall, annual water has cut by 15% between 1971 and 2000, considerably in southern and southeastern Morocco. By 2020, average annual rain is projected to decrease by 4% compared to 2000 levels, a development that may result in cereal yields falling between 10% fraction in conventional years and

FIGURE 9.1 Irrigation development projects in Upper Egypt, Egypt (Hamada, 2018).

by 50% in dry years (WHO, 2006). World climate change put together can increase seasonal variability and extremes leading to extra flooding. Between 20–30% advantage of the government's budget is spent on water management comes, like irrigation and water pipes and results are impressive: In 2005, 56% take advantage of the population in rural Southeast Mediterranean Sea countries had access to potable compared to 15% all through 1995 (World Bank, 2008a). However, per capita, water handiness is foretold to be reduced by 50% among the following 40 years (World Bank, 2008b), which may reverse this risk progress (World Bank, 2008b).

9.2 SCOPE OF THE CHAPTER

Particular attention ought to lean to some 500 million people who board countries that are terribly in danger of world climate change as a result of the physical location of their homes and social circumstances. Figure 9.2 below illustrates average rain in Africa and also the least developed countries (LDCs), that countries are usually the poorest, and that board LDCs that are at risk of quite one variety of weather disaster, i.e., floods, droughts, and storms; what is more as gradual environmental degradation like water level rise or natural action. The 10 most vulnerable countries are Comoros, Somalia, Burundi, Yemen, Niger, Eritrea, Afghanistan, Ethiopia, Chad, and Ruanda. These 10 most vulnerable nations have powerful nearly 180 storms or floods throughout the last 30 years. In these same countries, 11 million were tormented by drought in 2008 alone whereas 85 million are stricken by droughts in the last 30 years (CRED, 2008).

Adaptation will reduce overall vulnerability, specifically among the world's poorest. This might be through policies of finance in early warning and evacuation systems to organize people for storms, or serving to farmers to switch the crops grown and conjointly the temporal order of planting and harvest. The great news is that there are some success stories of poor countries reducing vulnerability to the impacts of world climate change. Bangladesh, one all told the country's most naturally in danger of world climate change, has taken steps over the past few years to become higher prepared, and thus, less vulnerable. These steps helped reduce mortality in the People's Republic of Bangladesh throughout Cyclone Sidr in 2007 that killed some 40 times fewer people than an identical scale cyclone in 1991 (3,400 deaths versus 138,000) that is despite the next population will increase over the intervening period.

9.3 POVERTY INCREASES VULNERABILITY TO CLIMATE CHANGE

Climate change is a serious danger to poverty reduction and threatens to undo a long time of development efforts. As the Johannesburg Declaration on Sustainable Development states, "the adverse outcomes of weather change are already evident, natural disasters are greater common and greater devastating and developing countries greater vulnerable." While

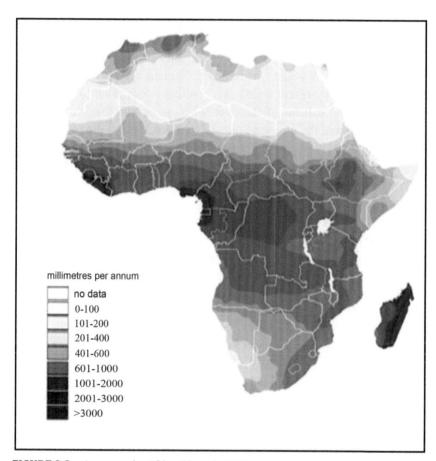

FIGURE 9.2   Average rain, Africa (Hamada, 2018).

climate change is a global phenomenon, its negative impacts are greater severely felt by poor humans and poor countries. They are extra vulnerable due to their excessive dependence on natural resources, and their limited capacity to address climate variability and extremes. Experience indicates that the best way to deal with climate change influences on the poor is through integrating adaptation responses into development planning.

Despite global efforts, poverty has become more widespread in many nations inside the last decade, making poverty discount the core undertaking for development in the 21st century. In the Millennium Declaration, 189 countries have resolved to halve extreme poverty via 2015. However, climate change is a serious risk to poverty discount and threatens to undo many years of development efforts. Persistently excessive tiers of starvation and malnutrition – 793 million chronically hungry people inside the world in 2014–2016 – and unsustainable human activity on the earth's carrying capability present a first-rate challenge for agriculture. To meet the growing food call for of the over 9 billion people that will exist by means of 2050 and the expected dietary changes, agriculture will want to supply 60% more food globally inside the same period. At the equal time, roughly one-third of food produced – 1.3 billion tons per year – is lost or wasted globally for the duration of the supply chain, with enormous economic and environmental costs (FAO 2019).

A placing link exists between growth in agriculture and the eradication of starvation and poverty. Agriculture widely understood – crop and live-stock production, fisheries, and forestry – provides income, jobs, food and other goods and services for almost all of humans now residing in poverty. As a result, basic GDP growth originating in agriculture is, on average, at least twice as effective in reducing poverty as growth generated in non-agriculture sectors, and as much as five instances greater effective than different sectors in resource poor low-earnings countries.

The cutting-edge trajectory of growth in agricultural production is unsustainable due to its negative impacts on natural resources and the environment. One-third of farmland is degraded, up to 75% of crop genetic variety has been lost and 22% of animal breeds are at threat. More than 1/2 of fish shares are completely exploited and, over the past decade, some thirteen million hectares of forests a year were transformed into different land uses.

The overarching challenges being faced are the growing scarcity and rapid degradation of natural sources, at a time when the demand for food, feed, fiber and items and offerings from agriculture (inclusive of crops, livestock, forestry, fisheries and aquaculture) is growing rapidly. Some of the highest population growth is expected in areas that are dependent on agriculture and already have excessive rates of meals insecurity.

Additional factors – many interrelated – complicate the situation: Competition over natural resources will retain to intensify. This might also come from urban growth, competition among various agricultural sectors, enlargement of agriculture at the cost of forests, industrial use of water, or leisure use of land. In many places this is a main to exclusion of traditional users from access to sources and markets; while agriculture is a first-rate contributor to climate change; it is also a victim of its effects. Climate change reduces the resilience of production systems and contributes to natural useful resource degradation. Temperature increases, modified precipitation regimes and severe climate events are expected to become extensively more severe within the future;

Increasing movement of people and goods, environmental changes, and modifications in production practices give rise to new threats from diseases (which includes incredibly pathogenic avian influenza) or invasive species (inclusive of nephritis fruit flies), which can affect food safety, human fitness and the effectiveness and sustainability of production structures. Threats are compounded via inadequate rules and technical capacities that could put complete food chains at danger; the policy schedule and mechanisms for production and resource conservation are on the whole disjointed. There is not any clear integrated management of ecosystems and/or landscapes.

9.4   POOR COUNTRIES SUFFER THE VAST MAJORITY OF HUMAN IMPACT OF CLIMATE CHANGE

To illustrate however erratically the human impact of world climate change is distributed, some basic statistics show on the burden of disasters that 98% of those affected in disasters between 2000 and 2004 and 99% of disaster casualties in 2008 were in developing countries (Watkins, 2007), International Strategy for Disaster Reduction Statistics (2008). Unequal access to property insurance is another example of people with low incomes is extra exposed to the impacts of world climate change. But 3% of the insured property losses, 1998–2007 values based on Munich Münchener Rückversicherungs-Gesellschaft (2009), from disasters are in low and lower-middle financial gain countries. Low-income households think about their biggest risk to be the incapacitation of the most earner, the CGAP organization on tiny insurance on behalf of the International Labor Organization (ILO) defines breadwinners because the members of households World Health Organization earn all or most of the financial gain, which implies that a disaster constitutes one in all the big risks they encounter. Figure 9.3 shows charting our water future, the Nile in Upper Egypt.

FIGURE 9.3   Charting our water future, the Nile in Upper Egypt, Egypt (Hamada, 2018).

Gradual environmental degradation as a result of world climate change, like dynamic rain patterns, put together disproportionately hurts the poor. Usually this can be often considerably the case once crop yields are reduced and farmers are forced to vary to extra drought resistant crops that supply less gain. The trendy GCF document to the UNFCCC Conference of the Parties (COP) informs that the GCF Board accredited USD 5.2 billion to aid the implementation of 111 weather change variation and mitigation tasks and packages in 99 growing countries. These projects and applications are predicted to attract USD 13.5 billion in direct public and personal sector co-financing. Of the 111 initiatives and programs authorized, 70% involve least developed countries (LDCs), Small Island growing States (SIDS) and African states. It is predicted that together these initiatives and programs will abate 1.5 billion metric tons of carbon dioxide equivalent (tCO_2e). The total value of the projects and programs accepted is USD 18.7 billion, International Strategy for Disaster Reduction Statistics (2019).

9.5 CLIMATE CHANGE EXACERBATES EXISTING INEQUALITIES FACED BY VULNERABLE GROUPS PARTICULARLY WOMEN, CHILDREN AND ELDERLY

The consequences of world climate change and impoverishment do not appear to be distributed uniformly among communities. Individual and social factors like gender, age, education, ethnicity, geographies, and language result in differential vulnerability and capability to adapt to the results of world climate change. World climate change effects like hunger, impoverishment, and diseases like diarrhea and malaria infection, dispro-portionately impact kids, i.e., regarding 90% of protozoan infection and symptom deaths are among young kids. What are exceedingly, in times of hardship, young girls are considerably potential to be taken out of school to care for sick relatives or earn an extra gain. The aged have weakened immune systems making them extra in danger of diseases and dynamic climatic conditions, notably heat waves, in conjunction with being very in danger of weather-related disasters as a result of reduced ability to move. Roughly 60% of cyclone Katrina victims were 65 years or older (Off, 2008).

9.6 IT IS A MATTER OF SOCIAL JUSTICE

It is a matter of social justice. If we have got a bent to worry regarding injustice and distinction, we have got a bent to ought to care regarding world climate change. World climate change exacerbates existing inequal-ities along with the varied vulnerabilities of men and girls. Climate change exacerbates gender in equalities. Women account for two-thirds of the world's poor (Irish Aid, 2009). 70 to 80% of agricultural staff are women, they're conjointly mostly to blame for water assortment and infrequently function the first caretaker in an exceedingly unit (IUCN, 2004). As a result, world climate change impacts like cut farm yields and water dispro-portionately impact women by reducing their livelihoods, impairing food provision, and increasing their organization employment. 75% of deaths in climate disasters are feminine (Aguilar, 2007), as a result of factors like an inability to swim that ends up in drowning throughout floods; constricting dress-codes inhibiting fast movement; and behavioral restrictions forbid-ding women from leaving the house without male relatives (Oxfam Canada,

2009). Women are also extra in danger of climate connected displacement and conflict, with women representing the bulk of people displaced of climate (Gender Action, 2009). These vulnerabilities and difference typically relegates women to the worst paid, least regulated jobs and may place them at higher risk of sexual exploitation (Gender Action, 2009).

9.7   A GLOBAL JUSTICE CONCERN: THOSE WHO SUFFER MOST FROM CLIMATE CHANGE HAVE DONE THE LEAST TO CAUSE IT

The world pollutants that contribute to global climate change do not adhere to national or regional boundaries. They impact people despite wherever they were produced and by whom. The worldwide poorest countries haven't benefited from the decades of the economic process that have accelerated warming. The poor are only accountable for a small part of the emissions that contribute to world climate change—nevertheless, they suffer the majority of the impact of climate change. These completely different realities of responsibility and human impact of world climate change raise necessary problems with world justice. Carbon dioxide gas emissions are believed to excessively the main reason of heating (Bernstein et al., 2007). Figure 9.4 Illustrates that agriculture is the world's biggest employer, Upper Egypt, Egypt.

FIGURE 9.4   Agriculture is the world's biggest employer, Upper Egypt (Hamada, 2018).

9.8 DIFFICULT CLIMATE JUSTICE ISSUES: HIGH GROWTH, DEFORESTATION, AND BLACK SOOT

Developed nations bear the foremost responsibility for world climate change; however, there's an increasing range of cases wherever low and middle financial gain countries conjointly contribute considerably to climate change. The highest 20 emitters of carbon dioxide enclosed giant and apace industrializing nations like China, India, Republic of Korea, Mexico, South Africa, Indonesia, and Brazil millennium development goals (MDGs) Indicators Statistics (2009). These countries typically have affluent natural resources and are experiencing fast economic development. Sometimes they notice it hard to achieve property policies as they're doing not invariably have access to applicable and affordable technologies. These are countries where climate justice issues became considerably acute but put together sensitive—being every giant emitter and very in danger of world climate change. Deforestation is another activity that raises the necessary climate justice issues. Whereas fuel usage is that the most important single contributor to world carbon dioxide emissions manufacturing world climate change (coal alone accounts for roughly 20% of world emissions) (US EPA, 2009; NEAA, 2009; Pew, 2009), deforestation put together plays a serious role, accounting for over 25% of world emissions (Howden, 2007). A majority of deforestation is done by cutting and burning (54%) and conjointly the rest constitutes of cattle ranching (5%), heavy logging (19%) and conjointly the growing palm oil industry (22%), a manufacture projected to grow to its use in biofuel production, Biofuels are created from biological materials like corn and are completely different from fossil fuels like oil made of long-dead biological substances (Hance, 2008). Barely sooner or later, deforestation generates the utmost quantity of greenhouse gas as 8 millions of people flying from London to New York (Howden, 2007).

As of 2003, 2 billion tons of carbon dioxide gas were joined to deforestation activities that drove to the destruction of 50 million acres—a zone roughly the scale of England, Wales, and Scotland (Howden, 2007). Speedy deforestation prioritizes immediate economic output in favor of natural wealth preservation. This has conjointly been the case historically: 500 years ago, nearly half the US, three-quarters of Canada and also the majority of Europe were wooded (University of Michigan, 2006). The majority of remaining forests globally are settled in high growth nations like Brazil, Indonesia, and China and conjointly in developing countries just like the Democratic Republic of Congo, some 57% of forest cover is found in developing nations, whereas 43% is in developed countries according to FAO, UNEP, and UNFF (2009): Forest loss, United Nations (UN) system-wide earth watch (http://earthwatchunep.ch/emerging issues/forests/).

In 2004, nearly 1%, Greenpeace UK (2009) states that up to 75% of Brazil's emissions return only from deforestation and in step with MDGs Indicators Statistic (2009), in 2004 Brazil emitted 1.22% of world carbonic acid gas emissions. Therefore, emissions from deforestation in Brazil equal some 0.92% of total emissions, of world carbon emissions were generated by clearing and burning the Amazon rain forest Greenpeace UK (2009), deforestation depletes natural resources always and leaves the land exposed to environmental disasters, inclusive with those related to climate change.

Poverty could also be a driver of practices that contribute to world climate change. Black carbon from the soot emitted from preparation stoves is one example. Under-ventilated fireplaces and primitive preparation appliances not only have negative health impacts nearly completely born by women, from smoke inhalation and respiratory sicknesses, but however, conjointly hurt the environment. Whereas carbon dioxide gas is that, the number one cause of world climate change—in charge for regarding 40% of warming—black carbon from soot is fast rising as an oversized contributor to climate change, inflicting as 18% of warming (Ramanatham, 2007). These findings are thus recent that they weren't lined among the 2007 Intergovernmental Panel on world climate change report. Even so, soot from fireplaces in tens of thousands of villages in developing countries is that the first contributor to black carbon, Ramanatham's analysis on Black Carbon is supported by Shindell and Faluvegi's study disclosed in March, 2009 in Nature Geoscience. Sometimes those that rely upon these preparation stoves to organize staple foods do not have access to affordable alternatives. A solution of this hard state of affairs would even have a world interest. Providing affordable alternatives might have a fast impact on curbing global warming, as in contrast to carbon dioxide gas that lingers among the atmosphere for years, soot solely remains for variety of weeks Viscount Nelson (2009).

9.9   COST OF HUMANITARIAN RELIEF IS EXPECTED TO GROW EXPONENTIALLY IN THE NEXT 20 YEARS

Climate change jointly threatens the flexibility of the international community to deliver humanitarian relief. The money needs for humanitarian facilitate are projected to increase by up to 1600% over succeeding 20 years, as a result of world climate change (Webster et al., 2008). Already these days, the funds getable for disaster ready and disaster relief are inadequate. Bilateral funds for disaster relief quantity $10 billion per annum (Watkins, 2007), are deed many disasters with little or no support, solely an awfully little proportion of world humanitarian facilitate goes into disaster state while usually often a crucial and worthy investment. Some specialists estimate that for every greenback invested in disaster preparation, six greenbacks can be saved in reconstruction costs (UNEP, 2004). Ultimately, the flexibility of individual households to protect themselves against the physical and economic shocks of disaster is the best avenue to assure survival.

9.10   COSTS OF ADAPTATION TO CLIMATE CHANGE

Climate adaptation refers to an individual or governmental action to reduce gift adverse effects or future risks of world climate change. This activity goes to be important to addressing the human impact of world climate change among the long run. However, to date, the investments are very restricted.

9.11   ADAPTATION NEEDS TO BE SCALED UP 100 TIMES TO AVERT WORST OUTCOMES

Developing nations have recognized the enormity of the climate change challenge, but the commitments to put money into funds in climate adaptation in developing countries amount to little or no. The triangular funds that are pledged for world climate change adaptation across developing countries presently amount to concerning $400 million. This quantity may be a smaller amount than the German state of Baden-Württemberg is getting to pay on strengthening flood defense (Watkins, 2007). The funds required for adaptation in developing countries substitute sharp distinction to this level of commitment. Specialists and Aid agencies estimate it's being worth of adaptation in developing countries ranges from 4 to $86 billion annually, UNFCCC (2005) estimates US$28–67 billion in 2030; Oxfam estimates its US$50 billion; International Bank for Reconstruction and Development (IBRD) estimates its US$9–41 billion in developing countries today; Stern estimates its US$4–37 billion in developing countries, a 110 Forum 2009: world climate change – The Anatomy of a Silent Crisis today; UNFCCC (2005) estimates it's $86 billion in 2015.

This vary depends on the world Bank estimate but includes 43 billion to adapt world climate change programs to accommodate climate change impacts what is more as two billion for disaster response every year; Extrapolating from UNFCCC worth estimates, funding needed for immediate 'climate-proofing' is between US$1.1 billion and US$2.2 billion for LDCs, rising to US$7.7–33 billion for all developing countries, with a median of $32 billion annually (Watkins, 2007; WRI, 2009), The African cluster, comprising over 50 nations, has calculable that $67 billion is needed annually from 2020 forayer for adaptation efforts in developing countries like building stronger defenses against rising ocean levels and developing drought resistant crops (Doyle, 2009).

It is vital to note that adaptation funding estimates a supplemental to existing overseas development aid wishes related to broader sustainable development and mitigation efforts. As an example, African nations more project that $200 billion annually is important to curb rising greenhouse gases by up energy potency and shift to renewable energy sources each year from 2020 forever (Doyle, 2009). Whereas these expenses are high, the worth of adaptation is way less than the worth of inaction. The Stern Report evaluated the value of ignoring world climate change at over than that of the two World Wars and conjointly the economic crisis, or 5–20% of GDP (Stern et al., 2006).

There are varieties of cases that supply a glimmer of hope. The alignment framework convention on amendment global climate change's national adaptation programs of action provides a method for least developed countries to identify priority world climate change adaptation activities (UNFCCC, 2005). Samoa has been hailed for his or her 2005 pertain and implementation efforts—expected to value $2 million, geared toward reducing vulnerability to increase resilience through shut collaboration with native communities (Jackson, 2008). As over 70% of

Samoa's population and infrastructure are settled in low-lying coastal zones, Samoan focus areas include: coastal ecosystems, coastal protection, community facility, forests, health and climate connected diseases, early warning systems, agriculture, and disaster risk reduction.

Bangladesh is an example of a state that has success invested with in disaster preparation to reduce the detrimental impacts of climate-connected disasters. It's among the countries most naturally in danger of world climate change but varied steps are confiscate the past few years to become higher prepared, and thus, less vulnerable. Cyclone Sidr hit the high lying, densely haunted coastal areas of People's Republic of Bangladesh in 2007, but disaster preparation measures like early warning systems and storm-proof homes cut the death individuals to 3,400 and restricted the economic damages to $1.6 billion (Bangladesh Government, 2008). Compared, the extremely inhabited delta region of the Ayeyarwady River in Myanmar wasn't ready for Cyclone Nargis in 2008 and also the human consequences were over 40 times larger—146,000 of us died, over two million of us became homeless and damages equaled around $4 billion (Relief, 2008). In August 2005, cyclone Katrina caught many in urban city suddenly and caused hurt among the reach of $100 billion (Nicholls et al., 2007; Wilbanks et al., 2007).

9.12   EXTREME WEATHER EVENTS WILL BOOST COSTS

The costs of forceful weather events are growing chop-chop. Since 1960, the amount of world weather-related disasters has exaggerated four-fold, real economic losses seven-fold and losses twelve-fold. Real losses are calculable to possess up from US $3.9 billion p.a. among the 50s to US $40 billion p.a. among the 90s. A component of the rise in disaster losses could also be explained by the explosive growth in human population, inappropriate land-use developing with (such as building on floodplains or areas in danger of abrasion or coastal storms), and the increasing cash value of homes and infrastructure and also the handiness of insurance. But world climate change and worsening weather extremes like winds, floods, and droughts could also be expected to play employment. Recent history has shown that weather connected losses can overwhelm insurance firms, which may respond by hiking premiums and retiring coverage from vulnerable sectors and regions. This might end in a higher demand for public-funded compensation and relief. Developing countries are most in danger of natural disasters. For many of them, weather-connected risks might become uninsurable, premium expenses might rise significantly, or insurance would become nonexistent or more durable to induce. Countries already grueling ironed to supply the requirements of food; safe water and shelter have little or no leeway for attention-grabbing the extra costs of natural disasters. Internal migrations will attainable increase the complete costs (Töpfer, 2003).

9.13 THE SEA LEVEL WILL RISE AS OCEAN WATERS WARM OBSERVED CHANGES

The sea level will raise as ocean waters warm observed changes:

- Enormous expanses of the oceans have hot over the past 50 years; globally, sea-surface temperatures have up in line with land temperatures.
- The worldwide mean water level has up by 10–20 cm throughout the 20th century—10 times faster than the estimate for the previous 3,000 years.
- Extra water is evaporating from the ocean surface; this has attainable resulted in total region vapor increasing by many percentages per decade over many regions of the northern hemisphere.
- 70% of sandy shorelines have withdrawer over the past 100 years; 20–30% is stable, whereas however 10% are advancing.
- Water is oozing into the fresh aquifers and intrusive into estuaries in low-lying coastal areas around the world, considerably on low-lying islands (Töpfer, 2003).

9.14 NEW RAINFALL PATTERNS WILL THREATEN WATER SUPPLIES

Observed changes: In Africa's giant catchment basins of Niger, Chad, and Senegal, total getable water has cut by 40–60%. Desertification operation has been exacerbated by lower average annual rain, runoff of the water and soil condition, notably in southern, northern, and western Africa (Töpfer, 2003).

The twenty-first century: rain patterns can still amendment round the world. Computer models consistently project that, as warming progresses, the temperate regions what are more as the geographic region will receive extra precipitation. In general, heating must accelerate the hydrological cycle. Hotter air causes extra water to evaporate. A warmer atmosphere can hold extra vapor, thus extra water is getable to fall back to Earth once it rains or snows. As a result, extreme precipitation events must become extra frequent and intense, leading to worse flooding. The Rhine floods of 1996 and 1997, the Chinese floods of 1998, the East European floods of 1998 and 2002, and also the Mozambique and European floods of 2000 all-purpose to a changing hydrological regime. Meanwhile, Central Asia, the Mediterranean region, the Sahel, and many of the various regions in Africa, Australia, and New Zealand are expected to receive considerably less rain. In addition, exaggerated evaporation in these regions may end up in drier conditions, with the following probability of drought. In many countries the implications of less precipitation and extra evaporation are going to be, larger stress on fresh provides. In addition, countries that depend upon run-off from mountains might suffer as glaciers retreat and snow accumulation reduces. Water shortages might have a control on critically important food production. Conflicts over water, considerably in stream and lake basins shared by over one country, may perhaps step up. Besides dynamic the distribution of precipitation, world climate change conjointly can have, an impact on the level of freshwater provides. Alga and plants grow extra prolifically in hotter conditions; when they decompose; higher levels of nutrients collect in the water. Meanwhile, extra intense rains will flush extra pollutants from the encircling land and from overflowing waste facilities. In regions wherever rainfall declines, pollutants are going to be additional extremely very targeted among the remaining getable water. Water quality goes to be put together tortured by lowland rise. Extra salty water will notice its approach into coastal aquifers and estuaries, creating fresh salt and eventually unsafe. This will have severe impacts in some areas, considerably low-lying islands, and atolls that believe underground water for his or her freshwater provides. Water intrusion conjointly can have a control on the surface fresh provides of communities living among estuaries. With one-third of the world's population living in countries that already lack enough water, and with populations and demand set to grow dramatically, fresh provides might even be one of our greatest vulnerabilities throughout a world climate change world (Töpfer, 2003).

9.15   WHAT CAN BE DONE?

The damage caused by extreme events could also be reduced through careful and consistent designing. Useful tools vary from effective land-use getting to municipal codes mandating that buildings be designed to resist high winds and ground subsidence to comprehensive coastal-management ways to early-warning systems like those already enforced in many hurricane-prone areas. The insurance business can contribute by seeking inventive solutions for reducing risks, therefore keeping sum of money offered and reasonable. Developing countries would like additional in depth access to insurance. Technology transfer and therefore the widespread introduction of micro-financing schemes and development banking might also facilitate (Töpfer, 2003).

Of course, every natural and social system will adapt impromptu to some extent. Such adaptation, however, will not be tight for many regions and sectors. And even planned adaptation will not realize addressing all impacts. Some distinctive and vulnerable natural and social systems (such as autochthonous communities) might even be irreparably stricken if the climate changes on the way aspect certain. Risks related to extreme weather events and unlikely-but potential large-scale singular events, just like the collapse of the West Antarctic ice sheet or the termination of the supposed Gulf Stream might even be considerably hardtop retort to. A heavy challenge facing these days is that there are still many uncertainties regarding world climate change impacts and our selections for adapting to them. There are simply too many variables – like population increment, the economy, technology, and environmental stress – that, a bit like the climate, will amendment over time.

Over succeeding century, East Africa might receive extra rain whereas Southern Africa will presumably become a decent deal drier. Food and water shortages are attainable to increase throughout most of Africa, as can floods and storms. Desertification process will keep a heavy threat in arid and dry regions.

9.16 CONCLUSION

This chapter provides a brief overview of background of potential alleviate poverty in rural areas for climate change adaptation, where the influences of climate change in developing nations can be dramatic, especially in regards to marginal areas and fantastically vulnerable poor communities. While GHG emission and global warming is basically a result of the activities of the industrialized countries, it's miles now a well-documented truth that the ramifications could be largely felt in developing countries, wherein the livelihoods of the poor are significantly at risk. Whilst the international community, and industrialized countries have identified the need to act, both to reduce GHG emissions and help growing countries adapt, the real mechanisms and help offered via the global community to this point have not come close to matching up to the needs and risks involved. In particular problems such decreasing vulnerability and assisting the poorest and maximum marginalized groups to adapt have been almost absent from the climate change agenda. Instead mechanisms have been in large part directed at addressing the issue of the mitigation of GHG emissions. Whilst this is manifestly a vital environmental solution, it does now not address the pressing issue of climate change and poverty.

It is, however, important to note that the Kyoto Protocol does have a commitment to sustainable development and will hence address the issue of poverty remedy. But is the Kyoto Protocol's implementing arm on sustainable development, the CDM, in reality capable of tackling this objective? This isn't the case for several reasons. Firstly, host countries will in my view outline their investments' improvement criteria, thus, opposition for investments by means of down-scaling improvement criteria is possibly. Secondly, for the reason that CDM initiatives involve in part the same risks as other FDI, they may be possibly to observe the same sample and go away the poorest countries out. Thirdly, this shows that the opting-out of the United States and the inclusion of large "hot air" income lead to higher competition among developing countries and to much less CDM investments, in particular in small scale projects.

The highest capability of the CDM to address together poverty alleviation and GHG abatement - the win-win initiatives - are precisely small-scale, off-gird tasks in micro-hydro and biomass energy generation. The potential of developing nations to evaluate projects, the inclusion of the poor within the design of CDM projects and streamlined approval techniques for small projects ought to all contribute to growing the CDM's effect on poverty. To suggest that ODA might be used to lessen those transaction expenses of in any other case commercially sustainable win-win projects. However, the overall assessment of the CDM's affects shows that these win-win projects are likely to be isolated successes, while the overall scale and nature of the investments are very far from what could be required to redress the inequity.

Whilst spotting those bilateral and multilateral donors should subsidize the CDM, overlaying transaction costs to encourage investment in smaller projects in poorer countries, the CDM does no longer have the capacity or the budget to fully cope with poverty and any pro-poor projects would only have isolated impacts. In contrast, the scale of investment and operations involved in ODA is substantial, allowing it to address poverty on a global scale. Any ODA subsidies to the CDM ought to therefore no longer be perceived as its contribution to the climate change agenda, however instead a small element of a larger multi-faceted attempt to address the vulnerability of the poor in growing countries.

KEYWORDS

- **agricultural examination systems**
- **agricultural productivity**
- **agriculture sector**
- **crop-yield losses**
- **drought-stress tolerant**
- **world climate change**

REFERENCES

Aguilar, L., (2007). Gender differences in deaths from natural disasters are directly linked to women's social and economic rights, Women's manifesto on climate change, UN commission on the status of women, 51st Session. *Emerging Issues Panel: Gender Perspectives on Climate Change.*

Bangladesh Government, (2008). *Cyclone SIDR in Bangladesh Damage, Lose and Needs Assessment for Recovery and Reconstruction.* Government of Bangladesh. http://

www.preventionweb.net/files/2275_CycloneSidrinBangladeshExecutiveSummary.pdf (accessed on 4 January 2020).

Bernstein, L., et al., (2007). *Climate Change 2007: Synthesis Report Summary for Policymakers.* Fourth assessment report of the intergovernmental panel on climate change.

CRED, (2008). *Center for Research on the Epidemiology of Disasters.* Collaborating center for research on the epidemiology of disasters school of public health, Catholic University of Louvain, Clos Chappelle Champs, Brussels, Belgium.

De Pledge, J., (2002). In: *Climate Change in Focus: the IPCC Third Assessment Report* (p. 19). Technical report, Royal Institute of International Affairs.

Doyle, A., (2009). *Africa Says Poor Need Billions to Fight Climate Fight.* Reuters. http://www.reuters.com/article/environmentNews/idUSTRE53J2RG20090420 (accessed on 4 January 2020).

Esper, J., et al., (2007). *Long-Term Drought Severity Variations in Morocco.* Swiss Federal Research Institute WSL, 8903 Birmsdorf, Switzerland Institute of Geography and NCCR Climate, University of Bern.

FAO (2019). Sustainable Development Goals, Sustainable agriculture, http://www.fao.org/sustainable-development-goals/background/fao-and-the-post-2015-development-agenda/en/.

FAO, UNEP, UNFF, (2009). *United Nations Forum on Forests, 2009.* Vital forest graphics. Nairobi, UNEP and Rome, FAO.

Friedman, T. L. (2008). Hot, Flat and Crowded –Why We Need a Green Revolution - And How it Can Renew America. Farrar, Straus & Giroux, p.158.

Gender Action, (2009). *Doubling the Damage: World Bank Climate Investment Funds Undermine Climate and Gender Justice.* Gender action. http://www.genderaction.org/images/2009.02_Doubling%20Damage_AR.pdf (accessed on 4 January 2020).

Global Deforestation. (2006). Global Deforestation. The University of Michigan, Lecture on January 4. http://www.globalchange.umich.edu/globalchange2/current/lectures/deforest/deforest.html (accessed on 4 January 2020).

Greenpeace UK, (2009). *Deforestation and Climate Change.* Greenpeace UK. http://www.greenpeace.org.uk/forests/climate-change (accessed on 4 January 2020).

Hamada, Y. M. (2018). Special pictures album, Took by Youssef M. Hamada, Egypt, 2018.

Hance, J., (2008). *Tropical Deforestation is One of the Worst Crises Since we Came Out of Our Caves.* Mongabay, http://news.mongabay.com/2008/0515-hance_myers.html (accessed on 4 January 2020).

Howden, D., (2007). *Deforestation: The Hidden Cause of Global Warming.* UK Independent, http://www.independent.co.uk/environment/climate-change/deforestation-the-hidden-cause-of-global-warming- 448734.html (accessed on 4 January 2020).

Irish, A., (2009). *What is Gender Equality?* Irish Aid Volunteering and Information Centre. http://www.irishaid.gov.ie/Uploads/Gender%20Inequality%20flyer.pdf (accessed on 4 January 2020).

IUCN (2004). *World Conservation Union - IUCN*, Diversity makes the difference, Actions to guarantee gender equity in the application of the Convention on Biological Diversity, Lorena Aguilar Revelo, 2004.

Jackson, C., (2008). *Samoa Praised Over Climate Change Plans*. The New Zealand Herald. http://www.nzherald.co.nz/climate-change/news/article.cfm?c_id=26&objectid= 10547372 (accessed on 4 January 2020).

McCarthy, M. (2008). "Why Canada Is the Best Haven From Climate Change." UK, Independent. July 4. http://www.independent.co.uk/environment/climatechange/why-canada-is-the-best-haven-from-climatechange-860001.html (accessed on 4 January 2020).

Millennium Development Goals Indicators Statistics (2009). The Millennium Development Goals Report, http://mdgs.un.org.

Münchener, R. G., (2009), *Königinstrasse 107*(8082). München, Germany. www.munichre.com (accessed on 4 January 2020).

NEAA, (2009). *Global Carbon Dioxide Emissions, 1970–2001*. Netherlands Environmental Assessment Agency. http://www.mnp.nl/mnc/i-en-0166.html (accessed on 4 January 2020).

Nelson, B., (2009). *Black Carbon Reductions Could Reverse Arctic Warming Within Weeks*. Eco Worldly. http://ecoworldly.com/2009/04/09/blackcarbon-reductions-could-reverse-arctic-warming within-weeks/ (accessed on 4 January 2020).

Nicholls, R. J., et al., (2007). Coastal systems and low-lying areas. In: Parry, M. L., et al., (eds.), *Climate Change 2007: Impacts, Adaptations and Vulnerability: Contribution of Working Group II to the Fourth Assessment Report of the Intergovernmental Panel on Climate Change*. Cambridge University Press, Cambridge, UK, Chapter 6 Box 6.4.

Off, G., (2008). *A Look at the Victims of Hurricane Katrina*. Scripps Howard, News Service. http://www.scrippsnews.com/node/30868 (accessed on 4 January 2020).

Oxfam Canada, (2009). *Climate Change*. Oxfam Canada. http://www.oxfam.ca/what-we-do/themesand-issues/climate-change/#_edn1 (accessed on 4 January 2020).

Pew, (2009). *Coal and Climate Change Facts*. Pew center on global climate change. http://www.pewclimate.org/global-warmingbasics/coalfacts.cfm (accessed on 4 January 2020).

Ramanatham, V., (2007). *Reduction of Air Pollution and Global Warming by Cooking with Renewable Sources*. Project Surya, Scripps Institution of Oceanography and Sri Ramachandra Medical College and Research Institute. http://www-ramanathan.ucsd.edu/SuryaWhitePaper.pdf (accessed on 4 January 2020).

Reduction Statistics, (2008). *International Strategy for Disaster and Centre for Research on the Epidemiology of Disasters Database*.

Relief Web, (2008). *Post-Margi's Joint Assessment*. Relief Web. http://www.reliefweb.int/rw/rwb.nsf/db900SID/ASAZ-7GRH55?OpenDocument (accessed on 4 January 2020).

Rocket Scientist, (2007). *Climate Change and Agriculture Country Note*. The Former Yugoslav, Republic of Macedonia. www.worldbank.org/eca/climateandagriculture (accessed on 4 January 2020).

Rosenthal, E., (2009). *Third-World Stove Soot is Target in Climate Fight*. The New York Times. http://www.nytimes.com/2009/04/16/science/earth/16degrees.html?_r=3&ref=world (accessed on 4 January 2020).

Stern, N., et al., (2006). *Stern Review: The Economics of Climate Change*. HM Treasury, Chapter 6. http://www.hm-treasury.gov.uk/stern_review_report.htm (accessed on 4 January 2020).

Töpfer, K., (2003). *UNEP, Information Unit for Conventions*. United Nations Environment, Program, CH-1219 Geneva, Switzerland; Web: www.unep.ch/conventions (accessed on 4 January 2020).

UNEP, (2004). *Extreme Weather Losses Soar to Record High for Insurance Industry.* UNEP. http://www.unep.org/Documents.Multilingual/Default.Print.asp?DocumentID= 414&ArticleID=4682&l=en (accessed on 4 January 2020).

UNFCCC, (2005). *National Adaptation Program of Action Samoa*. National Adaptation Program of Action Task Team (NTT), Ministry of Natural Resources, environment and meteorology, UNDP and GEF. http://unfccc.int/resource/docs/napa/sam01.pdf (accessed on 4 January 2020).

UNIDO, FAO, and IFAD, (2008). *The Importance of Agro-Industry for Socio-Economic Development and Poverty Reduction*. New York: UN Commission on Sustainable Development.

US EPA, (2009). *Human-Related Sources and Sinks of Carbon Dioxide*. US Environmental Protection Agency. http://www.epa.gov/climatechange/emissions/co2_human.html (accessed on 4 January 2020).

Watkins, K., (2007). *Human Develop Report 2007/2008 Fighting Climate Change: Human Solidarity in a Divided World*. United Nations Development Program, New York.

Webster M., et al., (2008). *The Humanitarian Costs of Climate Change* (p. 19). Feinstein International Center.

WHO, (2006). *Climate Change and its Impact on Health in Morocco*. World Health Organization, Eastern Mediterranean Regional Office.

Wilbanks, T. J., et al., (2007). Industry, settlement and society. In: Parry, M. L., et al., (eds.), *Climate Change 2007: Impacts, Adaptations and Vulnerability: Contribution of Working Group II to the Fourth Assessment Report of the Intergovernmental Panel on Climate Change*. Cambridge University Press, Cambridge, UK, Chapter 7 Box 7.4.

World Bank, (2007). *World Development Report 2008: Agriculture for Development*. Washington, D.C.: World Bank.

World Bank, (2008a). *Making the Most of Scarcity: Accountability for Better Water Management Results in the Middle East and North Africa*. World Bank. http://web. worldbank.org/WBSITE/EXTERNAL/COUNTRIES/MENAEXT/MOROCCOEXTN/ 0,contentMDK:21722173~menuPK:50003484~pagePK:2865066~piPK:2865079~theSi tePK:294540,00.html (accessed on 4 January 2020).

World Bank, (2008b). *Estimates That Per Capita Water Availability will Fall by Half by 2050 in the Middle East and North Africa*.

WRI, (2009). *Annual Adaptation Costs in Developing Countries*. World Resources Institute. http://www.wri.org/chart/annual-adaptation-costs-developingcountries (accessed on 4 January 2020).

CHAPTER 10

Future of Sugarcane Industry in Upper Egypt as a Case Study of Agribusiness

ABSTRACT

The poorest countries too would require additional capability thus on lure investment public and private. Emissions transfer schemes, just like the Clean Growth Mechanism, still profit within the main rising economies, wherever the necessity for brand new jobs, technology and investment isn't as nice. These schemes should not be joined to national economic interests, and company provides chains. Such connections create effectively another style of tied aid. It's to be untied and continue untied with relevancy economic and alternate interests. Copenhagen options a transparent order to fill the insufficiency in additional funding, providing for adaptation, along with reliable facilitate to those countries worst affected. Thus, on arranging the worldwide economy towards a low-carbon path, Copenhagen is perhaps going to get some kind of world system on emissions. It ought to opt for mechanisms and sanctions, along with a globally accepted answer on taxing CO_2. It's imperative, however, that the implications of such a system do not manufacture yet one more burden for the poor. What is going on to effectively act as a worldwide worth on carbon, can act as a regressive tax, like another taxes, since the additional costs of pollution will eventually be passed on to shoppers. The exaggerated costs will have the simplest result on the world's poorest teams, where individuals should be compelled to forgo an even bigger proportion of their gain otherwise pay on basic nutrition and health needs. Any climate policy ought to boot build amends for these effects through financial distribution, or risk a lot of intensifying inequities another time (Fust, 2009). During this context, this chapter critically studies the existence and sustainability of water collaboration efforts within the Nile basin as the river faces challenges from the universal weather change and shifting regional geopolitics.

As professional in agricultural economics and specializing in providing solutions to the problems of the agricultural economy through the creation of mathematical models, it was my duty in the tenth chapter providing a solution to the seriousness position of the water situation in Egypt, especially in this period in Upper Egypt as a case study of agribusiness. This problem of future of sugar industry (FOSI) and water scarcity model in Upper Egypt for maximize revenue of sugarcane farms, maximize efficiency of sugarcane factories in every governorate to give high production of sugar, maximize efficiency of black honey factories in every governorate to give high production of black honey, maximize efficiency of sugarcane juice shops in every governorate, maximize efficiency of sugarcane seeds for planting sugarcane farms in every governorate, minimize sugarcane lost in farms in every governorate, maximize labor wages of cultivation sugarcane crop, minimize water uses in planting sugarcane crop in every governorate, minimize energy consumption in cultivation sugarcane crop and minimize CO_2 emission in cultivation sugarcane crop in Upper Egypt as a case study of agribusiness as described in the book. The result of the mathematical model was the possibility of adaptation in the agricultural sector with that phase, increased agricultural production to achieve self-sufficiency in sugar and work sugarcane factories in Upper Egypt 100%, the possibility of employing full agricultural employment in Upper Egypt.

10.1 BACKGROUND

We live in a worldwide village where everyone have a responsibility to safeguard our planet. Isn't it logical and equitable; therefore, to insist that those who pollute have a duty to clean up? Pollution by some affects us all. All folks must perceive that pollution options a cost, and this cost ought to be borne by the polluter. Least accountable for greenhouse gas emissions are the world's poorest communities that suffer most from global climate change. This can be often primarily unjust. If efforts to make a worldwide framework to handle global climate change are to succeed and endure them need to be support the principles of fairness and equity. Individuals everyplace merit climate justice. And everyone over people ought to arise and demand specifically that from their representatives. An honest and easily approach would facilitate agreement at the world organization Climate Conference in Copenhagen later. We tend to cannot afford to fail. Global climate change is be a truly world issue. Its impacts are indiscriminate and threaten us all. People everyplace merit to not suffering because of global climate change. People everyplace deserve a future for his or her youngsters. People everyplace need to have leaders who realize the courageousness to achieve an answer to the current crisis (Annan, 2009). Global climate change is usually a threat to peace and stability. There is not an element of the globe which is able to be proof against the security threat. Figure 10.1 shows agribusiness, breaking Sugarcane in Esna, Upper Egypt, Egypt.

10.2   SCOPE OF THE CHAPTER

Conflicts are sometimes very difficult with multiple inter-dependent causalities, sometimes referred as 'complex emergencies.' global climate change has the potential to exacerbate existing tensions or output new ones—serving as a threat multiplier. It's usually a catalyst for violent conflict and a threat to international security (Smith and Vivekananda, 2007; WBGU, 2008). The United Nations Council held it's first-ever dialogue on the impact of global climate change in 2007 (UN, 2007). The links between global climate change and security are the subject of various status reports since 2007 by leading security figures at the US, UK and also the EU (Gunn, 2007; UK Mission, 2007; EU, 2008). The G77 cluster of developing nations to boot considers global climate change to be a heavy security threat that's anticipated to hit developing nations significantly onerous (Pomeroy, 2007).

10.3   AGRO-INDUSTRY AND AGRIBUSINESS IN CONTEMPORARY SOUTHEAST MEDITERRANEAN SEA

At a continental level, the broader category of agribusiness, inclusive jointly upstream (input) activities and downstream process activities, similarly as distribution and enhancing, is calculable to account for regarding one-fifth of GDP for Sub-Saharan Africa and slightly below half value of

FIGURE 10.1   Agribusiness, breaking Sugarcane in Esna, Upper Egypt (Hamada, 2018).

manufacturing and services (Jaffee et al., 2003). For individual African countries where information are out there, the share of total manufacturing value another of the two primary agro-industrial subsectors (food and beverages, and tobacco) alone ranges from 17% in South African to 47% in Ethiopia (World Bank, 2009).

Though many of the economies of Sub-Saharan Africa have undergone substantial structural modification with the share of agriculture in gross domestic product declining from 41% in 1960 to 12% in 2008, in most cases this has not been within the middle of the emergence of a dynamic and heterogeneous manufacturing sector. The key an element of the increase in industry's share in overall gross domestic product (from 17% in 1960 to 33% in 2008) was accounted for by the extractive industries. In the meantime the share of producing exaggerated solely marginally from 8.7% to 15%, having stagnated or maybe declined since 1995. Also, the service sector extended speedily from 34% in 1981 to 55% in 2008.

10.4 MANUFACTURING, AGRO-INDUSTRY, AND AGRIBUSINESS

The size and structure of the manufacturing sector in Africa, and among manufacturing of the agro-industrial sector, clearly differs well between every sub-regions and individual countries, it must be noted that regional definitions disagree in step with provide. Where information on similar indicators is provided by two or lots of sources, reference is made to the information not being directly comparable. Information on regional economic communities (RECs) must be treated with caution as results of overlapping memberships. Thus, information on South African Development Community (SADC) and Common Market for Eastern and Southern Africa (COMESA) is very problematic as a results of the inclusion of South Africa, the foremost vital industrial power in continent and a major exporter of farm products and agribusiness products, that distorts the image in reference to smaller member countries like Republic of Botswana, Malawi, and Zambia. At a regional level there is a clear division at the share of Manufacturing Value Added (MVA) in gross domestic product between Southern and North Africa (15.9% and 12.7% of gross domestic product severally) and Central, East and West Africa (9.2%, 8.4%, and 7.5% respectively), with the relative share of producing in overall gross domestic product static or declining across all regions at the past decade. These mixture figures conceal high levels of differentiation among sub-regions, wherever industrialization as a proportion of total gross domestic product ranges from highs of 16–18% (South Africa, Cote d'Ivoire, Cameroon) to lows of 3–5% (Botswana, Gabon, Ethiopia) (World Bank, 2009).

10.5 SOURCES AND STRUCTURE OF AGRO-INDUSTRIAL DEMAND

Actual and potential demand for the products of African agro-industry is evolving apace, driven by a range of agents along with rising per capita incomes, market and trade liberalization, dynamic technologies, increasing urbanization, with attendant changes in cultural norms and consumption patterns. In general, the proportion of financial gain spent on food declines with increasing financial gain. However, the quantitative relation of food process to agricultural worth another rises with income, increasing from around 0.1 in Uganda and the Kingdom of Nepal to around 0.4 in countries like Brazil, Mexico, and Argentina (World Bank, 2007). The overall marketplace for agro-industrial product in Africa is usually divided into four primary segments: (a) classic food staples; (b) fashionable urban supply; (c) classic export commodities; (d) non-traditional exports. Each among Africa and globally, the most demand trend has been the shift far away from the consumption of undifferentiated staple crops and towards exaggerated consumption of fruit, vegetables, vegetable oils, fish, and meat and farm product as a proportion of total hot intakes. This has translated into a shift far from undifferentiated primary goods product in international trade, towards higher another product categories. Although high value and non-traditional agro industrial production for export provide dynamic and growing market opportunities for some African countries, the foremost necessary demand driver in desert continent's, and might keep, the domestic and regional (intra-African) market. Diao et al. (2007) estimate domestic markets and intra-African trade to represent quite three-quarters of total worth at a continental level, with domestic markets alone represent 80% of total worth in regions like Eastern Africa.

10.6 SOUTHEAST MEDITERRANEAN SEA: WHY AGRIBUSINESS?

Manufacturing has not play about a dynamic role at the economic development of Southeast Mediterranean Sea to the current purpose. There are pressing problems that decisions for a reorientation to support agribusiness and agro-industrial development: particularly, destitution reduction and also the accomplishment of the MDGs; and guaranteeing equitable patterns of growth, that will address the concentration of employment and livelihoods at the agricultural sector. An agribusiness development path involving larger productivity growth throughout the full agribusiness worth chain—covering farms, firms and distributors—represents a solid foundation for fast, inclusive economical process and destitution reduction (Yumkella et al., 2011).

10.7 AGRIBUSINESS, ECONOMIC GROWTH AND POVERTY REDUCTION

Alongside its role in stimulating economic development, agribusiness and agro-industrial has the potential to contribute well to poverty reduction and improved social outcomes and an accord is rising that agro-industries are a decisive a part of socially-inclusive, competitive development strategies (Wilkinson and Rocha, 2008). Proof of the link between growth and destitution reduction varies in step with country. Spectacular economic and industrial growth in China raised 475 million people out of poverty between 1990 and 2005, though huge pockets of poverty still exist in growth-oriented areas and rural communities because of structural rigidities. In Sub-Saharan Africa, despite strong growth in recent years, the figure of individuals living on but $1.25 each day increased by 93 million during the identical period (Montalvo and Ravallion, 2010; United Nations agency, 2007).

The accomplishment of the MDGs in Sub-Saharan Africa has been forced by two factors: firstly of all, most countries haven't met the required gross domestic product rate of growth to reach the MDG1 target. Secondly, labor absorption and employment intensity are low as a result of degree of growth in some capital-intensive extractive sectors. Agribusiness directly contributes to the accomplishment of three key MDG's, significantly reducing destitution and hunger (MDG1), empowering women (MDG3) and developing world partnerships for growth (MDG8).

Strong synergies exist between agribusiness, agricultural performance and destitution reduction for Africa (World Bank, 2007); economical agribusiness may stimulate agricultural growth and powerful linkages between agribusiness and smallholders can cut back rural poverty. A spotlight on worth addition in agribusiness is, therefore, central to existing strategies for economic diversification, structural transformation and technological upgrading of African economies. Such a concentration is going to initiate faster progress towards prosperity, by poignant the bulk of the continent's economic activities and by harnessing vital linkages between the key economic sectors. This 'people-oriented' strategy will improve welfare and living standards of the scene majority of Africans, everyone as producers and shoppers, and from the perspective of employment, it's going to be troublesome to lift people out of poverty through direct employment generation in manufacturing alone, even with high employment snap and productivity growth, until the low initial manufacturing base can increase in importance. The indirect impact of industrialization growth on employment, however, is probable to be quite vital, significantly at the context of backward and forward linkages to agriculture and services through agribusiness development. In social insurance administration, such linkages are probable to possess nice potential, despite being presented rather weak as a results of a restricted proportion of agricultural raw materials being processed and exported and since essential industry-related services won't but have developed.

On the demand side, food expenditure sometimes represents the foremost vital single item of family expenditure, rising to quite half total expenditure for poor households in some countries, and so the efficiency of post-harvest operations might be a serious determinant of the prices paid by the urban and rural poor for food, and then a significant consider family food security (Jaffee et al., 2003).

Agro-industrial development can contribute to improved health and food security for the poor by increasing the overall accessibility, selection and organic process worth of nutrient, and facultative food to be hold on as a reserve against times of shortage, making sure that ample food is accessible that essential nutrients are consumed throughout the year. On the availability side, agro-industrial development options an instantaneous impact on the livelihoods of the poor every through exaggerated employment in agro-industrial activities, and via exaggerated demand for primary agricultural manufacture. Though varied significantly by subsector and region, agro-industry, considerably in its initial stages of development, is relatively labor-intensive, providing a range of opportunities for self and wage employment, at the case of Thailand, as an example, Watanabe et al. (2009) counsel that for the period 1988–2000 not exclusively was the number of employees per worth another for agro-processing at on prime of the mean for manufacturing overall, the number of poor employees per worth another at the agro-processing business was well larger. The figures for nutrient (and the smaller wood and wood product sector) were quite double those of the common of the manufacturing business, implying that the agro-processing business, considerably the food business, tends to hire an even bigger variety of the poor than different manufacturing industries.

Agro-industrial activity in Africa is to boot usually distinguished by a high proportion of female employment, ranging from 50% to as high as 90%, these figures exclude metals and energy. It must be noted that, among some market sectors a minimum of, strong gender segmentation in production and method tends to consign women to lots of vulnerable forms of work (casual, temporary and seasonal), lower paid and lots of labor-intensive preparation and/or processing (UNIDO et al., 2008; Wilkinson and Rocha, 2008). As an example, the 'non-traditional export sector' (vegetables, fruit and fish products), that's presently the foremost dynamic in terms of exports from Sub-Saharan Africa. Similarly, the small-scale food procedure and job operations ubiquitous throughout torrential of the continent are sometimes operated predominantly by ladies; a study of small-scale urban agro-processing and job enterprises in Cameroon found that over 80% were managed by female, with men being gift virtually solely within the mechanical milling/grinding and meat preparation activities (Ferré et al., 1999). Indeed, Charmes notes that the gender bias apparent in many agro-processing activities may contribute to the estimation of every agro-industrial activity and female employment in national accounting, noting that a very high share of these activities are undertaken as secondary activities and are sometimes hidden behind subsistence agriculture (Charmes, 2000).

Alongside job creation, agro-industrial enterprises sometimes provide crucial inputs and services to the farm sector for those with no access to such inputs, causation productivity and merchandise quality enhancements and stimulating market evoked innovation through chains and networks, facilitating linkages and allowing domestic and export markets to become more reciprocally supportive (FAO, 2007). Agro-industry is to boot amongst the foremost accessible of economic activities-frequently undertaken at small-scale, with low initial price and technological barriers to entry. SMEs keep key actors at the largely informal commerce and method networks, that dominate food procuration in torrential of (newly) urban Africa, and have proved remarkably adjective and resilient at the face of a range of economic, institutional and infrastructural challenges (Muchnik, 2003; Sautier et al., 2006).

10.8   PROMOTING AGRIBUSINESS DEVELOPMENT IN AFRICA

Agribusiness and agro-industry have the potential to contribute to a range of economic and social development processes, along with exaggerated employment generation (particularly female employment), dividend generation, destitution reduction and enhancements nutrition, health and overall food security. The entire identical, there keep substantial barriers to completely developing the business enterprise potential evident across the continent. Many of the facultative conditions required for property business enterprise development are not specific to the globe (or to manufacturing in general), but apply to any or all sectors of the economy. These embraces stable social science climate, sensible public governance along with functioning regulatory institutions, enforceable industrial laws and property rights, and adequate infrastructure and basic services, along with transport, ICTs and utilities. Traditionally, they have to boot self-enclosed the existence of a relatively high-capacity, interventionist state with active resource allocation and demand management strategies (Yumkella et al., 2011). A variety of policies, institutions and services are lots of directly relevant to the business enterprise sector. These involve lay alia: building the specified industrial capabilities and capacities; upgrading technology and innovation in terms of product and processes; strengthening group action capacities at the sector of production efficiency and business linkages, and cross-border cooperation; building capability to vary agro-industrial products; participating in world, regional and native worth chains; up rural infrastructure and energy security; promulgating standardization and interior control regulates, and establishing associated enfranchisement bodies; promoting institutional services for business enterprise; and mobilizing public-private sector cooperation on agribusiness development. Increasing the dimensions and aggressiveness of Africa's factory farm sector is critical—for farmers, agro-industrial enterprises and industry-related services. Indeed, the key challenge for developing agribusiness in Africa is that the improvement of manufacturing capacities and capabilities to beat constraints related to the event of economical industrial enterprises capable of rival in international, regional and domestic markets (Yumkella et al., 2011).

Driven by globalization and economies of scale, the international marketplace for agribusiness merchandise is sometimes characterized as market with some powerful actors—mainly huge multinationals and retailers—seeking the foremost cost-efficient suppliers worldwide, and where cut-throat competition is current. This has LED to a growing concentration, with food method companies' group action backward towards agriculture and forward towards the retail sector, bypassing ancient markets where smallholders sell to native markets and traders. The international market is very competitive in terms of value and merchandise quality, requiring ICT property and often 'just-in-time' delivery, with resultant high wants in terms of provision efficiency. Exacting wants are placed on suppliers to satisfy conformity standards and specifications, demanded by shoppers in developed countries and more and more by the growing people in rising economies. This makes it powerful, although not possible, for African agro-industries to 'break in and move up' the worldwide worth chain (UNIDO, 2009). In distinction, the national, sub regional and regional market in Africa has many competitive advantages for African producers in terms of proximity of markets and similarity of consumer preferences. Fashionable supermarkets and stores are apace increasing in many African countries. A key challenge is that the involvements of little farmers at the agro-industry provide chain.

10.9 WATER SCARCITY AND COMPARATIVE ADVANTAGES

One common suggestion to achieve water security throughout a water scarce country is to import product that require water for his or her production, rather than generating them domestically (Allan, 2001; Hoekstra and adorned, 2005; Zimmer and Renault, 2003). This may cut backpressure on water resources and would cause domestic water savings which is able to be used for different functions. Wichelns (2004) showed that this can be often not invariably true once exclusively resource endowments are thought-about ignoring production technologies or likelihood costs of water and different limiting factors. Technological variations were the first provider of comparative advantage to be known by economic experts (1817). The Ricardian model assumes a two of countries (A and B), a two of product (X and Y), and one single case of production (labor). Distinctions in technology are shapely by variations at the amount of output which is able to be obtained from one unit of work. Underneath these assumptions, country A faces a comparative advantage at the production of good X if it's comparatively some more of productive of this good, that is, if the opportunity cost of good X in terms of good Y is lower in country A than in country B. Compared to self-sufficiency, world output will increase and every countries gain from trade if they export the good within which they include a comparative advantage. Variations in technology or issue productivity are not the only real supply of comparative advantage. Variations in resource endowments play employment as incontestable by the Huckster-Ohlin model. The criterion version of this model assumes two countries, two goods, and a two of production factors. It assumes similar technologies and preferences in every countries; completely totally different issue endowments; and quality of things between industries but not between countries. Four central theorems are usually derived supported these assumptions: The Huckster-Ohlin theorem states that a state tends to export the good that intensively uses the consider factor that country.

The Stopler-Samuelson theorem states that an increase at the relative price of one good increase the real return of the factor used intensively at the production of that good and reduces the real return of the other factor. The Rybczynski theorem states that a rise at the endowment of one factor raises quite proportionately the production of the good that uses that issue relatively a lot of intensively and reduces the assembly of the other good. The problem worth feat theorem states that trade in final product is sufficient to bring leveling of factor prices. Within the context of comparative advantage, we have a bent to follow Wichelns (2004) to elucidate the potential impacts on trade and welfare from enhancements in irrigation efficiency.

Under the essential assumption of two countries (A and B), two product (rice and cotton), and two factors (land and water), let us to contemplate 1st that both of countries are water-scarce (available water resources are 180,000 m³ and 90,000 m³ in country A and B, respectively) and have totally different production technologies. Country A contains a technology level to output 6t/ha of rice or 2t/ha of cotton. The obtainable technology in country B is lower; it permits outputting 4t/ha of rice or 1t/ha of cotton.

The irrigation water needs for rice and cotton in both countries are 18,000 m³/ha and 6,000 m³/ha, respectively. Underneath these assumptions, country A may favor to irrigate 10ha of rice to provide a most of 60t of rice or irrigate 30ha of cotton to provide a most of 60t of cotton or any linear combination of areas for the assembly of rice or cotton in keeping with its production technology and factor endowments. Similarly, country B may irrigate 5ha of rice to provide 20t of rice or 15ha of cotton to provide 15t of cotton. Note that the irrigation water endowment limits production in both countries. Country A is comparatively water abundant; it's twice, as much water as country B and has an absolute advantage within the production of both products (higher yields per hectare). This might propose that country A would have a feature within the production of rice, the water-intensive crop. However, this is often not the case. The chance cost of outputting 1t of rice in country A, in terms of cotton production, is beyond in country B one ton of cotton compared to 0.75 ton of cotton, respectively). Therefore, country B (the water-scarce country) contains a comparative advantage within the production of rice (the water-intensive crop) and country A contains a comparative advantage within the production of cotton, wherever the chance cost of outputting one ton of cotton is lower. Both countries would gain from trade if they export the good within which they need a comparative advantage. The terms of trade, expressed as a quantitative relation, describe what quantity rice are needed to get 1t of cotton and can lie between the chance costs of outputting cotton in both countries (between 1 and 1.33).

Wichelns (2004) enhanced this instance to point that as long as water is that the limiting issue, country B will have a comparative advantage in rice production, whether or not it is a larger water endowment than country A. He also shows the presence of comparative advantage even once every countries have a similar production technology (crop yields are 6t/ha of rice and a 2 of t/ha of cotton) but completely different resource constraints. Water is that the limiting consider country A (180,000 m³ compared to 600,000 m³) and land is that the limiting consider country B (30 ha compared to 40ha). Underneath these assumptions, the water-abundant country (country B) will have a comparative advantage at the assembly of the water-intensive crop (rice). Therefore, the chance costs are determined by the assembly coefficients, the water wants, and conjointly the inadequacy conditions. Within this context, let us to presently ponder an improvement in irrigation effectiveness that's translated into lower irrigation water wants. Suppose a decrease at the irrigation water wants for rice in country from 18,000m³ to 12,000m³ (all different assumptions keep the same). As water is that the limiting considers country A, the new technology permits irrigating lots of hectares with a similar amount of water resources. Country A may irrigate 15 ha of rice to provide 90 t of rice. As a result, the chance costs amendment in country A. The chance cost of outputting 1 t of rice is 0.66 t of cotton (lower than before) and conjointly the cost of outputting 1t of cotton is 1.5 t of rice (higher than before).

Considering the new chance costs in country A, the comparative advantages are reversed once each country face water inadequacy. Country A options a comparative advantage in rice production and country B in cotton production. Once every country has a similar production technology but completely different resource constraints the reduction at intervals the irrigation water demand is not strong enough to lower the possibility value of producing rice. Therefore, country B (water-abundant country) still options a comparative advantage in rice production (water intensive crop) and country A (water-scarce country) in cotton production (no water-intensive crop). Whereas many of the propositions of these theoretical models are lost by generalization or once considering lots of realistic assumptions (WTO, 2008), comparative advantage continues to predict and build a case for the gains of trade. Trade-focused calculated general equilibrium models (CGE) are, to some extent, empirical applications of these theories. They are supported by the classical (Walrasian) general equilibrium theory and incorporate a theoretical and coherent framework.

10.10   WATER SCARCITY AND THE IMPACT OF IMPROVED IRRIGATION MANAGEMENT: A COMPUTABLE GENERAL EQUILIBRIUM ANALYSIS

Water might be a scarce resource. About 40% of the world's population these days faces shortages in spite of whether or not they establish in dry areas or in areas wherever rainfall is torrential (Molden, 2007). The biggest user of sweet water is the agricultural sector—globally around 70% of all fresh are used for food production. However, but 60% of all the water used for irrigation is effectively consumed by crops. Consistent with the enhancements in irrigation management would be economically helpful the amount of water savings that may be achieved. Throughout the approaching decades, water inadequacy is predicted to rise due to a fast increase within the demand for water because of growth, urbanization, and a consumption of water per capita. By 2025, the world's population is anticipated to rise from 6.5 billion these days to 7 billion. Over 80% can live in developing countries and 58% in quickly growing urban areas (Rose Grant et al., 2002). Consequently, 1.8 billion people are expected to alive in countries or regions with absolute water inadequacy, and two-thirds of the world population are often at a beneath stress conditions (UN, Water/ FAO, 2007). Also, global climate change will influence the availability of water, modifying the regional distribution of freshwater resources (UN, Water/FAO, 2007).

According to the United Nations (2006), throughout the last century, irrigation water use has exaggerated doubles rapid as population, permitting the world food system to reply to the increasing growth in population. However, increasing irrigated areas won't be ample to confirm future food security and meet the increasing demand for water in densely populated settled however water scarce regions (Kamara and Sally, 2004). Therefore, a technique to handle the matter is to scale back the inefficiencies in irrigation. Seckler et al. (1998) rated that around 50% of the long run increase (by 2025) within the demand for water are often met by increasing irrigation efficient. Currently, irrigation efficient in most of the developing countries is acting poorly; the sole exception is water-scarce North Africa, wherever levels are appreciating those discovered in developed regions. Certainly, there are variations in performance at regions. Rose grant et al. (2002) imply that irrigation efficient ranges between 25% and 40% within the Philippines, Thailand, India, Pakistan, and Mexico; between 40% and 45% in Malaysia and Morocco; and between 50% and 60% in Taiwan, Israel, and Japan. Foremost developing regions that suffer from water inadequacy like the Middle East, North Africa, South Asia, and enormous parts of China and India, irrigated agriculture contributes considerably to total crop production. Simply the center East, North Africa, and South Asia account for around 43% of the full world water used for irrigation purposes. This part studies potential world water savings and its economic implications. Higher levels of irrigation efficient imply that the identical production might be achieved with less water (generating water savings) or, instead, that a lot of hectares might be irrigated by the identical obtainable water resources (implying higher production). Consequently, regional use of fresh resources and comparative benefits amendment is modifying regional trade patterns and welfare. The net impact on water use, therefore, depends on a fancy interaction between sectors and regions implying changes in provide and demand altogether sectors affected.

Improving irrigation potency worldwide generates new chance prices that may reverse regional comparative benefits in food production. Regions with comparatively poor irrigation performance might expertise positive impacts in food production and exports once up irrigation effectiveness. At the identical time, food-exporting regions are vulnerable to positive impacts evoked by increased irrigation potency elsewhere. International trade of food merchandise isn't solely the most channels through that welfare impacts unfold across regions, it's conjointly seen as a key variable in agricultural water management. As water becomes scarce, commerce product that need torrential water for his or her production might save water in water-scarce regions.

Most of the prevailing literature associated with irrigation water use investigates irrigation management, water productivity, and water use potency. One prevailing of literature compares the performance of irrigation systems and irrigation ways generally (Pereira, 1999; Pereira et al., 2002). Others have a transparent regional focus and target specific crop varieties. To produce some examples from this in depth prevailing: Deng et al. (2006) investigate enhancements in agricultural water use potency in arid and dry areas of China. Bluemling et al. (2007) study wheat-maize cropping pattern within the North China plain. Mailhol et al. (2004) analyze ways for hard wheat production in Tunisia. Lilienfeld and Asmild (2007) estimate excess water use in irrigated agriculture in western Kansas. As the on crack of examples indicates, water problems related to irrigation management are sometimes studied at the farm level, the river-catchment level or the country level. These studies omit the international dimension of water use. A full understanding of water use and conjointly the impact of improved irrigation management isn't doable whereas not understanding the international marketplace for food and connected product, like textiles. Figure 10.2 shows water, jobs and the economy in Upper Egypt, Egypt.

This is often what has been understood from the previous studies though not at the worldwide level. There is a trend to increase this approach by scrutiny the initial water savings with the maximum water savings taking into consideration adjustment processes in food and different markets. This is often a lot of attention-grabbing since it's very on the face of it that regions can amendment otherwise to the initial water savings. Describe potential impacts on trade and welfare from enhancements in irrigation efficiency supported the comparative advantage theory and the studies on economic models of water use. The last section at the chapter will presents a new model conjointly the information on water resources and water use.

10.11 FUTURE OF THE SUGARCANE INDUSTRY AND WATER SCARCITY IN UPPER EGYPT AS A CASE STUDY OF AGRIBUSINESS

The Arab Republic of Egypt is in two continents, Africa and the Sinai Peninsula in Asia. It's the 15th most populous settled country within the

FIGURE 10.2   Water, jobs and the economy, Upper Egypt, Egypt (Hamada, 2018).

world and also the second most populous in Africa after Nigeria. It has one million square kilometers (386.6 sq. miles) of land of that less than 4% is cultivated for crop production. The 99 million folks live on 40,000 sq. It's formed by three tributaries, the White Nile, the Blue Nile, and also the Atbara River. Agriculture may be a major economic issue in Egypt. It's a problem as an area food supply, for international trade, for balance of payments, land use and water use and as a basic product for food and fiber producing. Therefore each side of the economic structure of Egypt relates to agriculture. Banking, transportation, tax and tariff structure, subsidies, native and international markets and health are all an element of the agricultural system of a country, to not mention politics after all. Agriculture's contribution to Egypt's gross domestic product (GDP) is bit-by-bit decreasing, but it's still a very important activity. Albeit exclusively 3% of the whole space is tillable land, agriculture accounted for 13.4% of gross domestic product in financial 2005 and 27.1% of total employment in 2012. Egypt, due to its very limited cultivable land resource and water resources, is maybe more dependent on research to expand food production than any other country within the world. Moreover, the first beneficiary of such researches is the consumer, who is served by having not only an adequate supply of food, however, additionally higher quality and less expensive food further. Therefore, the requirement for a high-quality, productive agricultural research program is important to a sound economy and a stable political future. Whereas abundant has been achieved through past support of agricultural analysis, way more effort was required.

Sugar is one amongst the most substrates of the human diet. The five top sugar manufacturing countries within the world are India, Brazil, Thailand, Australia, and China. Their production accounts for 40% of the entire world sugar production out of the a hundred and fifteen countries manufacturing sugar within the world. Out of those countries, 67 countries is manufacturing sugar from sugarcane, 39 countries manufacturing it from sugar beet and 9 countries manufacturing it from each of sugarcane and sugar beet. Thus, 70% of the sugar is made from sugarcane and 30% from sugar beet in Egypt. Sugar business is one among the oldest industries in Egypt. The Egyptians were the first to reach the refined sugar industry throughout the ninth and tenth centuries, a part of this production was exported to Europe. Egypt still depends entirely on sugar cane until 1981 which was the main problem within the Egyptian sugar industry. There are eight sugar factories in Egypt distributed within the major governorates that manufacture sugar cane. To achieve equity and efficiency in sugar industry nonlinear programming model will be formulated to focus on the scientific linkage between water, equity and full capacity of sugar cane industries.

The filling of the Grand Ethiopian Renaissance Dam (GERD) reservoir is expected to start in June 2020, at the next rainy season. Egypt fears it is not able to face water shortages if that process isn't always achieved slowly. While Ethiopia wants to fill the reservoir within 4 years, Egypt desires a slower pace that can be numerous in response to droughts. Egypt also wishes Ethiopia to guarantee a minimum annual drift of 40bn cubic meters of water in non-drought situations and to keep water degrees in Egypt's Aswan dam above 165 meters. Actually, International experts say it ought to be viable to reach an agreement at the joint management of the river system and that there are examples from other river basins such as the 1960 Indus Waters Treaty among India and Pakistan. "Agreements for filling the Grand Ethiopian Renaissance Dam (GERD) should consider the possibility of droughts occurring at the filling process, which may include arrangements on how the three countries may adapt beneath these conditions," said Kevin Wheeler, of the Environmental Change Institute at Oxford University. "Considering shared drought management strategies over the long term—after the filling process is complete—is also very important." "We need a fair settlement and we understand Ethiopia's development needs," stated the Egyptian official. "But Ethiopia has to understand that we depend on the Nile for 97% of our water." Figure 10.3 shows ground water developments in Upper Egypt, Egypt.

FIGURE 10.3   Ground water developments, Upper Egypt, Egypt (Hamada, 2018).

10.12   FUTURE OF SUGARCANE INDUSTRY AND WATER SCARCITY IN UPPER EGYPT

Sugar industry is one among the oldest industries in Egypt. Pharaohs had been extracting sugar from the Kharroub. The Far East is considered the primary home for extracting sugar from sugarcane and it spread to Persia and also the Arabs who transferred it to Egypt in 710 AD. The Egyptians were the first to reach the refined sugar industry during the ninth and tenth centuries, a part of this production was exported to Europe. Sugar industry had entered the stage of collapse within the Mamluk era and returned to prosper another time within the era of Muhammad Ali (Hassan and Nasr, 2008).

In 1850, Egypt tended import good varieties of high productivity sugarcane seeds from the Far East, so increasing the Egyptian produc-tion of sugarcane and this encouraged to establish of the oldest sugar plant in Al-Rawda in El-Menia governorate (Al-Suefi, 1974), followed by establishing regarding 16 factories on the Upper Egypt within the era of Khedive Ismail to provide the raw sugar that was exported to Europe to be refined. In 1881, the first factory to refine sugar was established in Al Hawamdia town by the name The Egyptian refining company (Al-Suefi, 1974). A French shareholding company was established beneath the name of the sugar factories in Upper Egypt in 1892, and then the two firms merged during a single company on name of the General Company for sugar factories and refining in 1897.

After the revolution, specifically, in 1956 there was a partial nation-alization of the sugar industry and also the sugar company and also was established the Egyptian company distillation was established followed by the nationalization of the corporate and it became a completely public sector company closely-held by the state (Al-Suefi, 1974). Sugar produc-tion in Egypt still depends entirely on sugarcane until 1981 which was the main problem in the Egyptian sugar industry, because of the being of Egypt within the heart of the arid region and deficit of water resources additionally to the speedy increase of population which needs to continue the growth within the production of sugar to fill the native need, it abso-lutely was tough to expand the cultivation of sugarcane crop as a result of it needs massive quantities of water and is against the Egyptian water policy. This problem has necessitated adapting the sugar beet crop in Egypt. Egypt has allotted tracts of reclaimed land since 1981 for growing sugar beet and established the first factory for sugar beet in the Hamul in the Governorate of Kafr El-Sheikh in 1982, wherever it is participated about 2.5% total sugar production in Egypt (El-Deeb and Ibrahim, 1999).

Sugar industry in Egypt differs from the others, whether in South America or Africa, because the farmers grow sugarcane and beet in their own lands, and providing the factories with needs by supply contracts.

Whereas in the New World and a few African countries dominates a system of plantations that are owned by the corporate that owns the factory using machines and wide workforce further because of the magnitude of the farm which might in some cases thousands of hectares. The cultivation of sugarcane is focused within the southern governorates wherever Qena governorate comes within the first rank, that alone accounted for 48% of the total cultivated space of sugarcane and Assouan governorate comes in second place and accounted for 25% of the cultivated space of sugarcane in Egypt. El-Menia comes within the third rank with regarding 12% of the total cultivated space of sugarcane in Egypt, whereas Luxor comes within the fourth rank with regarding 7% of the total space and eventually, Sohag within the fifth place with regarding 5.4% of the total space planted with sugarcane in 2007 in Egypt. The main focus of cultivating sugarcane and sugar industry within the provinces of the Middle and Upper Egypt is principally as a result of the appropriate natural conditions for agriculture, that ultimately mirrored on higher average productivity per feddan (more than 51 tons/feddan). This productivity puts Egypt within the initial place among the manufacturing countries of sugarcane within the world. The common of productivity per feddan declines bit by bit from southern to northern Egypt (Hassan and Nasr, 2008). Egypt is thus considered the most important producer of sugarcane among Arab countries followed by Sudan, at 7.5 million tons annually (ESCWA, 2009).

There are eight sugar factories in Egypt distributed within the major governorates that produce sugarcane. The governorate with the most cultivated land of sugarcane own the most factories in Qena, and distributed in Nagy Hammadi – Deshna – Kous – Armant, whereas in Assouan, there are two factories in Komombo and Edfu. There's just one plant in each Sohag (Gerga) and El-Menia (Abou Korkas). The production capability of Nagy Hammadi is the first that comes at the primary rank, followed by Kous plant, Armant and Edfu. The sugar factories in Qena stands the first in sugar production as they share with regarding 56% of the sugarcane production in Egypt, whereas Assouan governorate comes second because it share with regarding 33% of sugar production. Egypt suffers from the shortage of the domestic production to cover the demand of sugar due to the rise of population and also the rise of the quality of living. Despite of the rise of Egyptian sugar production till it reached regarding 1.6 million tons in 2006, however, it produces just one-third of domestic demand of sugar. Egypt was self-sustaining of sugar till 1973 and it had a surplus for export once the production was more than the consumption. In 2006, the total consumption of sugar in Egypt was about 2.6 million tons, and there is a deficit, of one million tons, that can be compensated by the import from abroad (FAO, 2006).

The sugar is one among the backed commodities provided by the government at low costs for folks with limited income, and most important of those products bread flour, vegetable oil and sugar (Löfgren and El-Said, 2001). Egypt comes at the seventh rank among the sugar consumed nations within the world in 2007, its share reached to 1.7% of the total world consumption of sugar (Anonymous, 2008). There's additionally a relentless increase within the quantities of sugar consumed in Egypt that turned Egypt from an exported country at the start of the seventeenth to an imported country by regarding 2.6 million tons. The increasing demand of sugar consumption was mirrored on the rise of consumption per capita, where the sugar consumption per capita in 1972 was estimated with about 16 kg per year then increased to about 34 kg per year in 2007. With this average, Egypt comes beneath high level of consumption per capita among the African countries and worldly additionally as the global average of sugar consumption per capita was estimated around 22 kg per capita/year in 2007 (UN, 2007).

Egypt has achieved self-sufficiency of sugar till 1973 and it even had a surplus to be exported abroad, wherever sugar was one among the exported product that contribute to get foreign currencies as the cotton crop. Within the late of 1973, the deficit of sugar production started, because the native production couldn't meet the demand, that there have been a requirement to import from abroad to fill the deficit. Though Egypt is among Africa's largest sugar producers, its average annual consumption that ranges between 2.2 million tons exceeds native production. Egypt therefore imports massive amounts of sugar to make amends for the shortage in supply. The major reduction in the value of imports from 1998/1999 to 1999/2000 is primarily related to the major drop in sugar world market prices at the same year. An increase of sugar import was seen since 1973 where the gap started between production and consumption. Then, the imports continued to grow till 1990 (from 680 thousand tons of sugar to 455,000 tons in the following year) as a result of growth of sugar beet cultivation and also the operational of the second plant, leading to increasing domestic production of 830,000 tons in 1990 to regarding 1.31 million tons, that led to scale back imports of sugar. With the continued increase in cultivated space and increasing the capacities of sugar beet factories, the domestic production exaggerated and imports of sugar began to decline until it reached 401,000 tons in 2001. Since the year 2002, the imports began to extend once more due to the increasing of population and increase of sugar demand that reached regarding one million ton. Egypt imports quantities of sugar in the form of pure sugar to be refined regionally where two factories were established to refine sugar (the imported and the local product), the first in Al-hawamdia and also the second on the road between Cairo and Ismailia. Figure 10.4 Sugarcane transport to the factory in Upper Egypt, Egypt.

To establish Achieving efficiency and equity in sugar industry in Upper Egypt under scarce of water and climate change linkages between water uses efficiency and sugar factories work efficiency and equity in planting sugar crops varieties in old and new land of Upper Egypt to became agribusiness

FIGURE 10.4   Sugarcane transport to the factory, Upper Egypt, Egypt (Hamada, 2018).

as future of agriculture in Southeast Mediterranean Sea. The poverty maps in different parts of the Upper Egypt show distinct characteristics and a strong correlation to the agricultural systems and managed water access. Poverty level within the Nile Basin ranges from 17% in Egypt to over 50% in five of the Nile Basin countries (Awulachewa et al., 2012). This requires agricultural development strategies to aim for more productive use of water and to maximize the profit gained from the water consumed.

10.12.1 MATHEMATICAL MODEL OF FUTURE OF SUGARCANE INDUSTRY (FOSI)

Subject to

Future of sugar industry (FOSI) and water scarcity model in Upper Egypt for maximize revenue of sugarcane farms, maximize efficiency of sugarcane factories in every governorate for give high production of sugar, maximize efficiency of black honey factories in every governorate for give high production of black honey, maximize efficiency of sugarcane juice shops in every governorate, maximize efficiency of sugarcane seeds for planting sugarcane farms in every governorate, minimize sugarcane lost in farms in every governorate, maximize labor wages of cultivation sugarcane crop, minimize water uses in planting sugarcane crop in every governorate, minimize energy consumption in cultivation sugarcane crop and minimize CO_2 emission in cultivation sugarcane crop in Upper Egypt as a case study of agribusiness can be written as follows:

Maximize $\sum \sum \sum \sum \sum \sum \sum \sum \sum \sum \sum \sum \sum \sum \sum \sum \sum \sum \sum \sum \sum \sum \sum \sum \sum \sum \sum \sum$

$[(A_{\text{Future-of-sugar-industry-in-Upper-Egypt}} - B_{\text{Future-of-sugar-industry-in-Upper-Egypt}}) * C_{\text{Future-of-sugar-industry-in-Upper-Egypt}}]$
$+ [((D_{\text{Future-of-sugar-industry-in-Upper-Egypt}} * C_{\text{Future-of-sugar-industry-in-Upper-Egypt}}) / (E_{\text{Future-of-sugar-industry-in-Upper-Egypt}})$
$* 100] +] + [((D_{\text{Future-of-sugar-industry-in-Upper-Egypt}} * C_{\text{Future-of-sugar-industry-in-Upper-Egypt}}) / (F_{\text{Future-of-sugar-industry-in-Upper-Egypt}}) * 100]] + [((D_{\text{Future-of-sugar-industry-in-Upper-Egypt}} * C_{\text{Future-of-sugar-industry-in-Upper-Egypt}}) / (G_{\text{Future-of-sugar-industry-in-Upper-Egypt}}) * 100] + [((D_{\text{Future-of-sugar-industry-in-Upper-Egypt}} * C_{\text{Future-of-sugar-industry-in-Upper-Egypt}})$
$/ (H_{\text{Future-of-sugar-industry-in-Upper-Egypt}}) * 100]$ (1)

Subject to

$\sum \sum \sum \sum \sum \sum \sum \sum \sum \sum \sum \sum \sum \sum \sum \sum \sum \sum \sum \sum \sum \sum \sum \sum \sum \sum \sum \sum$

$A_{\text{Future-of-sugar-industry-in-Upper-Egypt}} \geq MAX\_A_f \text{ for } A_f$ (2)

$$\sum \sum \sum \sum \sum \sum \sum \sum \sum \sum \sum \sum \sum \sum \sum \sum \sum \sum \sum \sum \sum \sum \sum \sum \sum \sum$$
$$B_{\text{Future-of-sugar-industry-in-Upper-Egypt}} \leq \text{MIN\_B}_f \text{ for } B_f \tag{3}$$

$$\sum \sum \sum \sum \sum \sum \sum \sum \sum \sum \sum \sum \sum \sum \sum \sum \sum \sum \sum \sum \sum \sum \sum \sum \sum \sum$$
$$C_{\text{Future-of-sugar-industry-in-Upper-Egypt}} \geq \text{MAX\_C}_f \text{ for } C_f \tag{4}$$

$$\sum \sum \sum \sum \sum \sum \sum \sum \sum \sum \sum \sum \sum \sum \sum \sum \sum \sum \sum \sum \sum \sum \sum \sum \sum \sum$$
$$D_{\text{Future-of-sugar-industry-in-Upper-Egypt}} \geq \text{MAX\_ D}_f \text{ for } D_f \tag{5}$$

$$\sum \sum \sum \sum \sum \sum \sum \sum \sum \sum \sum \sum \sum \sum \sum \sum \sum \sum \sum \sum \sum \sum \sum \sum \sum \sum$$
$$E_{\text{Future-of-sugar-industry-in-Upper-Egypt}} \geq \text{MAX\_ E}_f \text{ for } E_f \tag{6}$$

$$\sum \sum \sum \sum \sum \sum \sum \sum \sum \sum \sum \sum \sum \sum \sum \sum \sum \sum \sum \sum \sum \sum \sum \sum \sum \sum$$
$$F_{\text{Future-of-sugar-industry-in-Upper-Egypt}} \geq \text{MAX\_F}_f \text{ for } F_f \tag{7}$$

$$\sum \sum \sum \sum \sum \sum \sum \sum \sum \sum \sum \sum \sum \sum \sum \sum \sum \sum \sum \sum \sum \sum \sum \sum \sum \sum$$
$$G_{\text{Future-of-sugar-industry-in-Upper-Egypt}} \geq \text{MAX\_ G}_f \text{ for } G_f \tag{8}$$

$$\sum \sum \sum \sum \sum \sum \sum \sum \sum \sum \sum \sum \sum \sum \sum \sum \sum \sum \sum \sum \sum \sum \sum \sum \sum \sum$$
$$H_{\text{Future-of-sugar-industry-in-Upper-Egypt}} \geq \text{MAX\_ H}_f \text{ for } H_f \tag{9}$$

MAX_A_f Maximum revenue for planting sugar crop variety f (10)
MIN_B_f Minimum total costs for planting sugar crop variety f (11)
MAX_C_f Maximum land area available for planting sugar crop variety f (12)
MAX_D_f Maximum sugar crop yield in planting sugar crop variety f (13)
MAX_E_f Maximum working sugarcane factory efficiency for sugar crop variety f (14)
MAX_F_f Maximum working black honey factory efficiency for sugar crop variety f (15)
MAX_G_f Maximum working juice shops efficiency for sugar crop variety f (16)
MAX_H_f Maximum sugarcane variety seeds for planting sugarcane farms f (17)
MIN_I_f Maximum main sugar crops price in planting sugar crop variety f (18)
MIN_J_f Maximum secondary sugar crops price in planting sugar crop variety f (19)
MIN_K_f Maximum labor wages cost in planting sugar crop variety f (20)
MIN_L_f Minimum draft animals cost in planting sugar crop variety f (21)
MIN_M_f Minimum machinery cost in planting sugar crop variety f (22)
MIN_N_f Minimum seeds cost in planting sugar crop variety f (23)
MIN_O_f Minimum manure coast in planting sugar crop variety f (24)
MIN_P_f Minimum fertilizers coast in planting sugar crop variety f (25)
MIN_Q_f Minimum insecticides coast in planting sugar crop variety f (26)
MIN_R_f Minimum laser land leveling cost in planting sugar crop variety f (27)
MAX_S_f Maximum other expenses coast in planting sugar crop variety f (28)
MIN_T_f Minimum rent coast in planting sugar crop variety f (29)
MIN_U_f Minimum sugar crops water consumption in planting sugar crop variety f (30)
MIN_V_f Minimum energy uses in cultivation in planting sugar crop variety f (31)
MIN_V_f Minimum CO_2 emission in cultivation in planting sugar crop variety f (32)

Variables

A _{Future-of-sugar-industry-in-Upper-Egypt} Estimated revenue for planting main crop yield f and secondary crop yield u of sugar crop variety t of sugar crop u in sub-sugar crop-group r in sub-soil type e (Old land or New land) in sub-zone (Upper Egypt) o in sub-season f by total crop production cost s include labor wages cost u draft animals cost g machinery cost a seeds cost r manure coast i fertilizers coast n insecticides coast d laser land leveling cost u rent cost s and other expenses coast t by total energy consumptions r include energy consumption for irrigation y energy consumption for labor i energy consumption for draft animal n energy consumption for land preparation u energy consumption for seed planting p energy consumption for manure p energy consumption for fertilization e energy consumption for insecticide r energy consumption for laser land leveling e and energy consumption for other expenses g by total kerosene fuel consumption y crop water consumption p and crop emission t.

B _{Future-of-sugarcane-industry-in-Egypt} Estimated total costs of planting main crop yield f and secondary crop yield u of sugar crop variety t of sugar crop u in sub-sugar crop-group r in sub-soil type e (Old land or New land) in sub-zone (Upper Egypt) o in sub-season f by total crop production cost s include labor wages cost u draft animals cost g machinery cost a seeds cost r manure coast i fertilizers coast n insecticides coast d laser land leveling cost u rent cost s and other expenses coast t by total energy consumptions r include energy consumption for irrigation y energy consumption for labor i energy consumption for draft animal n energy consumption for land preparation u energy consumption for seed planting p energy consumption for manure p energy consumption for fertilization e energy consumption for insecticide r energy consumption for laser land leveling e and energy consumption for other expenses g by total kerosene fuel consumption y crop water consumption p and crop emission t.

C _{Future-of-sugarcane-industry-in-Egypt} Estimated land area allocated for planting main crop yield f and secondary crop yield u of sugar crop variety t of sugar crop u in sub-sugar crop-group r in sub-soil type e (Old land or New land) in sub-zone (Upper Egypt) o in sub-season f by total crop production cost s include labor wages cost u draft animals cost g machinery cost a seeds cost r manure coast i fertilizers coast n insecticides coast d laser land leveling cost u rent cost s and other expenses coast t by total energy consumptions r include energy consumption for irrigation y energy consumption for labor i energy consumption for draft animal n energy consumption for land preparation u energy consumption for seed planting p energy consumption for manure p energy consumption for fertilization e energy consumption for insecticide r energy consumption for laser land leveling e and energy consumption for other expenses g by total kerosene fuel consumption y crop water consumption p and crop emission t.

D $_{Future-of-sugarcane-industry-in-Egypt}$ Estimated sugar crop variety yield production allocated in planting main crop yield f and secondary crop yield u of sugar crop variety t of sugar crop u in sub-sugar crop-group r in sub-soil type e (Old land or New land) in sub-zone (Upper Egypt) o in sub-season f by total crop production cost s include labor wages cost u draft animals cost g machinery cost a seeds cost r manure coast i fertilizers coast n insecticides coast d laser land leveling cost u rent cost s and other expenses coast t by total energy consumptions r include energy consumption for irrigation y energy consumption for labor i energy consumption for draft animal n energy consumption for land preparation u energy consumption for seed planting p energy consumption for manure p energy consumption for fertilization e energy consumption for insecticide r energy consumption for laser land leveling e and energy consumption for other expenses g by total kerosene fuel consumption y crop water consumption p and crop emission t.

E $_{Future-of-sugarcane-industry-in-Egypt}$ Estimated sugarcane variety production, for working the sugarcane factory in this governorate in high efficiency and give high production of sugar, from main crop yield f and secondary crop yield u of sugar crop variety t of sugar crop u in sub-sugar crop-group r in sub-soil type e (Old land and New land) in sub-zone (Upper Egypt) o in sub-season f by total crop production cost s include labor wages cost u draft animals cost g machinery cost a seeds cost r manure coast i fertilizers coast n insecticides coast d laser land leveling cost u rent cost s and other expenses coast t by total energy consumptions r include energy consumption for irrigation y energy consumption for labor i energy consumption for draft animal n energy consumption for land preparation u energy consumption for seed planting p energy consumption for manure p energy consumption for fertilization e energy consumption for insecticide r energy consumption for laser land leveling e and energy consumption for other expenses g by total kerosene fuel consumption y crop water consumption p and crop emission t.

F $_{Future-of-sugarcane-industry-in-Egypt}$ Estimated sugarcane variety production, for working black honey factories in this governorate in high efficiency and give high production of black honey, from main crop yield f and secondary crop yield u of sugar crop variety t of sugar crop u in sub-sugar crop-group r in sub-soil type e (Old land and New land) in sub-zone (Upper Egypt) o in sub-season f by total crop production cost s include labor wages cost u draft animals cost g machinery cost a seeds cost r manure coast i fertilizers coast n insecticides coast d laser land leveling cost u rent cost s and other expenses coast t by total energy consumptions r include energy consumption for irrigation y energy consumption for labor i energy consumption for draft animal n energy consumption for land preparation u energy consumption for seed planting p energy consumption for manure p energy consumption for fertilization e energy consumption for insecticide r energy consumption for laser land leveling e and energy consumption for other expenses g by total kerosene fuel consumption y crop water consumption p and crop emission t.

G $_{Future-of-sugarcane-industry-in-Egypt}$ Estimated sugarcane variety production, for working juice shops in every governorate in high efficiency, from main crop yield f and secondary crop yield u of sugar crop variety t of sugar crop u in sub-sugar crop-group r in sub-soil type e (Old land and New land) in sub-zone (Upper Egypt) o in sub-season f by total crop production cost s include labor wages cost u draft animals cost g machinery cost a seeds cost r manure coast i fertilizers coast n insecticides coast d laser land leveling cost u rent cost s and other expenses coast t by total energy consumptions r include energy consumption for irrigation y energy consumption for labor i energy consumption for draft animal n energy consumption for land preparation u energy consumption for seed planting p energy consumption for manure p energy consumption for fertilization e energy consumption for insecticide r energy consumption for laser land leveling e and energy consumption for other expenses g by total kerosene fuel consumption y crop water consumption p and crop emission t.

H $_{Future-of-sugarcane-industry-in-Egypt}$ Estimated sugarcane variety seeds for planting sugarcane farms in every governorate, from main crop yield f and secondary crop yield u of sugar crop variety t of sugar crop u in sub-sugar crop-group r in sub-soil type e (Old land and New land) in sub-zone (Upper Egypt) o in sub-season f by total crop production cost s include labor wages cost u draft animals cost g machinery cost a seeds cost r manure coast i fertilizers coast n insecticides coast d laser land leveling cost u rent cost s and other expenses coast t by total energy consumptions r include energy consumption for irrigation y energy consumption for labor i energy consumption for draft animal n energy consumption for land preparation u energy consumption for seed planting p energy consumption for manure p energy consumption for fertilization e energy consumption for insecticide r energy consumption for laser land leveling e and energy consumption for other expenses g by total kerosene fuel consumption y crop water consumption p and crop emission t.

10.12.2 RESULTS AND DISCUSSION OF MODEL FUTURE OF SUGARCANE INDUSTRY (FOSI)

Future of sugar industry (FOSI) and water scarcity in Upper Egypt is a model formulated as an analytical tool to apply simulates future of sugar industry and water scarcity in six governorates of Upper Egypt in sugar fields after laser land leveling did in this governorates and in the season in the sector under water scarcity in Upper Egypt (Figure 10.5).

In addition, the model has the flexibility of introducing alleviate poverty system as a pre-requisite to achieve efficiency and equity in sugar fields in Upper Egypt under the global financial and climate change, reduce cost to become competitive in the world market, reduce water consumption and reduce social cost of pollutants on sugar fields. The economic, financial, risk, and the annual internal rate of return analysis of sugarcane varieties production are also investigated. Several steps were followed to implement FOSI model: first step was optimum sugar cropping pattern for each season for every governorate in old and new land of Upper Egypt, second step simulated optimum sugar cropping pattern for Upper Egypt, third step simulated optimum sugar cropping pattern for the all governorates with existing sugar crop-ping pattern (2011/2012–2015/2016) to reallocate sugar crops acreage according to efficiency use of water. To populate the model, field data reported by farmers was used. The data required was collected through a comprehensive survey of water consumption and other inputs to sugar crops fields on a seasonal basis, and included a comprehensive data set relating to the farm enterprise and associated socio-economic condi-tions. Sugarcane variety area, yield, and cost data were obtained from the MALR (2018). Data of water consumption were collected from the MWRI (2018). Necessary data pertaining to sugar cropping pattern input of the respective production system were collected from primary sources and converted into corresponding sugar cropping pattern values. Greenhouse gases emissions were calculated and represented per unit of the energy input. Data presented in this study are representative of typical and/or average data recorded over the five consecutive years of 2011/2012–2015/2016.

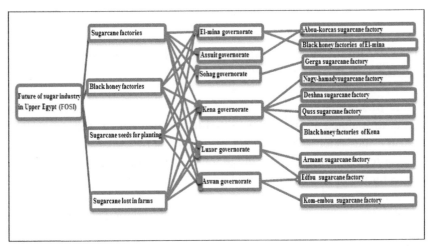

FIGURE 10.5   Structure model of future of sugar industry (FOSI) and water scarcity in Upper Egypt (FOSI, 2018).

10.12.2.1   OPTIMAL SOLUTIONS

Future of sugar industry (FOSI) and water scarcity in Upper Egypt is a model should be used. In order to suitable soil type and water could be reallocated to increase farm income, the model adjusted whatever change in land was needed to accompany the changes in soil type and water after made laser land leveling in old and new land of Nile valley in Upper Egypt. The model structure to optimal cultivation based on suitable soil type and water in Upper Egypt is given in Figure 10.5. Moreover, data in Tables 10.1 and 10.2 indicates that sugarcane variety areas of the considered scenario as a homogeneous character is higher than their heterogeneous one. In addition, the water requirement rates of the homogeneous consideration were less than the heterogeneous one.

TABLE 10.1   Changes of Sugarcane Varieties in Upper Egypt Flow Values from Mean (2011/2012–2015/2016) to (FOSI)*

	Sugar cane varieties in El-menia				Sugar cane varieties in Suhag				Sugar cane varieties in Qena			
	Mean	FOSI	Change	%	Mean	FOSI	Change	%	Mean	FOSI	Change	%
54(S9)	15118.740	8221.345	-6897.4	-45.6	6765.79	1693.640	-5072.2	-75.0	49176.62	1397.815	-47778.8	-97.2
PH 80/13	14.708	6577.076	6562.4	44616.5	2.856	0.000	-2.9	-100.0	4.63	0.000	-4.6	-100.0
G 84/47	0.445	526.166	525.7	118086.4	58.044	0.000	-58.0	-100.0	180.61	1048.362	867.8	480.5
EH 9/16	0.000	0.000	0.0	0.0	0.000	0.000	0.0	0.0	1.96	0.000	-2.0	-100.0
B 2/264	0.000	0.000	0.0	0.0	1.289	0.000	-1.3	-100.0	73.99	1397.815	1323.8	1789.3
B 2/266	0.000	0.000	0.0	0.0	0.000	0.000	0.0	0.0	0.00	0.000	0.0	0.0
G 96/74	0.000	0.000	0.0	0.0	0.000	0.000	0.0	0.0	0.00	0.000	0.0	0.0
Coba 8/203	0.462	312.411	311.9	67521.4	0.000	1767.277	1767.3	176727.7	0.00	24884.7	24884.7	2488468
G 99/103	0.000	345.296	345.3	34529.6	0.000	3902.737	3902.7	390273.7	0.00	0.000	0.0	0.0
S 57/14	0.000	0.000	0.0	0.0	0.000	0.000	0.0	0.0	0.00	3495	3495	349454
G 47/2003	0.000	0.000	0.0	0.0	0.000	0.000	0.0	0.0	0.00	12231	12231	1223089
G 49/2003	0.000	0.000	0.0	0.0	0.000	0.000	0.0	0.0	0.00	0.000	0.0	0.0
Other	26.846	460.395	433.5	1614.9	0.000	0.000	0.0	0.0	12.23	1397.815	1385.6	11325.1

	Sugar cane varieties in Luxor				Sugar cane varieties in Aswan				Sugar cane varieties in Upper Egypt			
	Mean	FOSI	Change	%	Mean	FOSI	Change	%	Mean	FOSI	Change	%
54(S9)	1132.69	1059.427	-73.3	-6.5	36606.44	3239.944	-33366.5	-91.1	108800	15612	-93188.12	-85.65
PH 80/13	0.00	0.000	0.0	0.0	0.00	0.000	0.0	0.0	22.197	6577.076	6554.88	29530.47
G 84/47	849.52	794.570	-54.9	-6.5	9.66	1079.981	1070.3	11079.9	1098.276	3449.079	2350.80	214.04
EH 9/16	0.00	0.000	0.0	0.0	0.00	0.000	0.0	0.0	1.961	0.000	-1.96	-100.00
B 2/264	0.00	0.000	0.0	0.0	13.10	2159.963	2146.9	16388.5	88.376	3557.778	3469.40	3925.71
B 2/266	0.00	0.000	0.0	0.0	1.31	1079.981	1078.7	82052.8	1.315	1079.981	1078.67	82052.84
G 96/74	8495.19	7945.701	-549.5	-6.5	0.00	0.000	0.0	0.0	8495.186	7945.701	-549.48	-6.47
Coba 8/203	6796.15	7680.844	884.7	13.0	0.84	4319.925	4319.1	514176.8	6797.451	38965.137	32167.69	473.23
G 99/103	0.00	0.000	0.0	0.0	0.00	0.000	0.0	0.0	0.000	4248.033	4248.03	424803.32
S 57/14	0.00	0.00	0.0	0.0	1.34	1079.98	1078.6	80761.1	1.336	4574.520	4573.18	342406.73
G 47/2003	9911.05	7945.701	-1965.3	-19.8	1.32	4319.925	4318.6	327464.8	9912.369	24496.512	14584.14	147.13
G 49/2003	0.00	0.000	0.0	0.0	0.00	0.000	0.0	0.0	0.000	0.000	0.00	0.00
Other	1132.69	1059.427	-73.3	-6.5	14.35	4319.925	4305.6	30009.9	1186.120	7237.563	6051.44	510.19

*Green is the values that have increased, Red is the values that have decreased.

Source: (1) MALR (2018) (2) ECAPMS (2018) (3) FOSI Model (2018).

Table 10.3 shows economic evaluations for optimum cultivation based on suitable soil type, laser land leveling in old and new land of Nile valley in Upper Egypt and water and it is comparison with existing condition of Upper Egypt. Figure 10.6 shows changes in crops area aggregates in the six governorates of Upper Egypt from mean 2011/2012–2015/2016 to FOSI in old and new land of Upper Egypt. The results show that area of sugar crops would be 117.744 thousand hectares cultivated in the six governorates of Upper Egypt, respectively and the proposed model provided higher net benefit than the existing model for all cases. The sum of net benefit for heterogeneous case (1833.612 million E.P.) was higher than the sum of homogeneous case (1376.478 million E.P.) in Upper Egypt and the sum of sugar crops water consumption for heterogeneous case (2213.366 million cub. M.) was lower than the sum of homogeneous case (4133.443 million cub. M.).

TABLE 10.2 Changes of Sugarcane Varieties in Upper Egypt Area, Unit Values and Aggregate Zones Flow Values From Mean (2011/2012–2015/2016) to (FOSI)*

	Sugar cane varieties in El-menia				Sugar cane varieties in Suhag			
	Mean	FOSI	Change	%	Mean	FOSI	Change	%
Irrigated area of sugar cane	15.161	16.443	1.28	8.45	6.828	7.364	0.54	7.85
Soil type	0.11	0.12	0.01	8.45	0.05	0.05	0.00	7.85
Main sugar crops yield	0.78	1.72	0.94	121.30	633.40	934.79	301.38	47.58
Sugar percent %	0.002	0.004	0.00	70.13	0.165	0.184	0.02	11.46
Main sugar crops price	12.09	15.66	3.57	29.50	5.45	7.01	1.57	28.77
Secondry sugar crops price	0.00	0.00	0.00	0.00	0.00	0.00	0.00	0.00
Total sugar crops production cost	215.39	332.19	116.81	54.23	100.07	148.77	48.70	48.66
*Labor Wages	34.23	91.77	57.53	168.07	25.21	41.10	15.89	63.03
Draft Animals	0.00	0.00	0.00	0.00	0.00	0.00	0.00	0.00
Machinery	24.01	43.22	19.22	80.05	15.61	19.36	3.75	24.02
Seeds Cost	0.00	10.02	10.02	1002.22	0.00	4.49	4.49	448.83
Manure	0.00	0.00	0.00	0.00	0.00	0.00	0.00	0.00
Fertilizers	37.98	55.20	17.23	45.36	14.14	24.72	10.58	74.78
Insecticides	0.00	0.00	0.00	0.00	0.00	0.00	0.00	0.00
Laser land leveling cost	1.68	2.09	0.40	23.95	0.76	0.94	0.18	23.25
Other Expenses	19.46	32.02	12.57	64.59	8.76	14.34	5.58	63.67
Rent	72.70	97.87	25.2	34.62	35.77	43.83	8.1	22.55
Sugar crops revenue	252.46	575.44	322.98	127.93	213.27	312.49	99.22	46.53
Sugar crops profit	132.60	243.25	110.65	83.45	111.64	163.72	52.08	46.65
Sugar crops water consumption	380.72	274.10	-106.62	-28.00	173.92	154.38	-19.54	-11.23
Kerosene fuel Liter	23.87	19.63	-4.25	-17.78	8.70	7.87	-0.83	-9.51
Energy consumption in cultivation	1276.87	1102.79	-174.08	-13.63	453.63	442.28	-11.34	-2.50

	Sugar cane varieties in Qena				Sugar cane varieties in Luxor				Sugar cane varieties in Aswan			
	Mean	FOSI	Change	%	Mean	FOSI	Change	%	Mean	FOSI	Change	%
Irrigated area of sugar cane	49.597	45.852	-3.75	-7.55	28.317	26.486	-1.83	-6.47	36.648	21.600	-15.05	-41.06
Soil type	0.35	0.33	-0.03	-7.28	0.20	0.19	-0.01	-6.47	0.26	0.15	-0.11	-41.06
Main sugar crops yield	1.27	5.71	4.44	349.18	0.89	2.28	1.40	157.13	0.00	2.20	2.20	91314.23
Sugar percent %	1.323	1.200	-0.12	-9.32	0.294	0.686	0.39	133.20	0.000	0.549	0.55	54.94
Main sugar crops price	39.44	43.67	4.23	10.71	22.59	25.22	2.64	11.68	29.23	20.57	-8.66	-29.63
Secondry sugar crops price	0.00	0.00	0.00	0.00	0.00	0.00	0.00	0.00	0.00	0.00	0.00	0.00
Total sugar crops production cost	793.28	845.40	52.12	6.57	454.27	554.27	100.00	22.01	587.92	455.61	-132.30	-22.50
*Labor Wages	174.96	246.07	71.11	40.65	100.19	144.03	43.84	43.76	129.67	109.54	-20.12	-15.52
Draft Animals	0.00	0.00	0.00	0.00	0.00	0.00	0.00	0.00	0.00	0.00	0.00	0.00
Machinery	114.44	118.36	3.92	3.43	65.53	87.02	21.49	32.79	84.81	79.92	-4.90	-5.77
Seeds Cost	44.51	49.24	4.73	10.63	20.97	22.83	1.86	8.87	27.14	26.64	-0.50	-1.83
Manure	0.12	0.15	0.04	29.81	0.07	0.00	-0.07	-100.00	0.09	0.00	-0.09	-100.00
Fertilizers	147.17	143.69	-3.46	-2.37	84.28	85.13	0.85	1.01	109.07	74.57	-34.50	-31.63
Insecticides	0.00	0.00	0.00	0.00	0.00	0.00	0.00	0.00	0.00	0.00	0.00	0.00
Laser land leveling cost	5.49	5.24	-0.25	-4.63	3.15	3.36	0.22	6.89	4.07	2.74	-1.33	-32.64
Other Expenses	75.71	81.29	5.58	7.37	43.35	54.23	10.88	25.10	56.11	46.49	-9.62	-17.14
Rent	238.77	250.70	11.9	5.00	136.73	157.65	20.9	15.30	176.98	115.71	-61.2	-34.61
Sugar crops revenue	1452.16	1908.98	456.82	31.46	300.45	763.72	463.27	154.19	981.62	736.14	-245.48	-25.01
Sugar crops profit	641.98	936.67	294.69	45.90	96.33	209.46	113.12	117.43	393.92	280.52	-113.40	-28.79
Sugar crops water consumption	1543.21	871.22	-671.99	-43.54	887.34	503.25	-384.03	-43.26	1148.32	410.41	-737.91	-64.26
Kerosene fuel Liter	81.80	55.36	-26.44	-32.33	39.26	32.55	-6.71	-17.08	52.48	26.55	-25.93	-49.41
Energy consumption in cultivation	4725.79	3110.43	-1615.35	-34.18	2109.00	1796.70	-312.31	-14.81	2729.48	1465.24	-1264.24	-46.32

Green is the values that have increased, Red is the values that have decreased.

Source: (1) MALR (2018) (2) ECAPMS (2018) (3) FOSI Model (2018).

It indicates that the variation of heterogeneous character had a large impact on the optimal solution. For this reason, FOSI model with heterogeneous character of land area was appropriate for finding cultivation based on suitable soil type after laser land leveling in old and new land of Nile valley in Upper Egypt have done.

TABLE 10.3   Changes of Economic and Financial Values in Sugarcane Varieties in Upper Egypt Zones From Mean (2011/2012–2015/2016) to (FOSI)*

	Sugar cane varieties in El-menia				Sugar cane varieties in Suhag				Sugar cane varieties in Qena			
	Mean	FOSI	Change	%	Mean	FOSI	Change	%	Mean	FOSI	Change	%
Irrigated area of sugar cane	15.161	16.443	1.28	8.45	6.828	7.364	0.54	7.85	49.597	45.852	-3.75	-7.55
Total sugar crops production cost	-215.39	-332.19	-547.58	-11680.9	-100.07	-148.77	-248.84	-4869.59	-793.28	-845.40	-1638.68	-5211.93
Sugar crops revenue	252.46	575.44	322.98	127.93	213.27	312.49	99.22	46.53	1452.16	1908.98	456.82	31.46
Sugar crops profit	132.60	243.25	110.65	83.45	111.64	163.72	52.08	46.65	641.98	936.67	294.69	45.90
Rate of return (IRR)	0.17	0.73	0.56	325.40	1.13	1.10	-0.03	-2.71	0.83	1.26	0.43	51.47
Absolute Risk	90.46%	39.69%	-50.77%	-56.13	32.90%	22.45%	-10.45%	-31.75	22.24%	16.92%	-5.32%	-23.93

					Sugar cane varieties in Aswan			
	Mean	FOSI	Change	%	Mean	FOSI	Change	%
Irrigated area of sugar cane	28.317	26.486	-1.83	-6.47	36.648	21.600	-15.05	-41.06
Total sugar crops production cost	-454.27	-554.27	-1008.53	-9999.8	-587.92	-455.61	-1043.53	13230.04
Sugar crops revenue	300.45	763.72	463.27	154.19	981.62	736.14	-245.48	-25.01
Sugar crops profit	96.33	209.46	113.12	117.43	393.92	280.52	-113.40	-28.79
Rate of return (IRR)	-0.34	0.38	0.72	-211.60	0.67	0.62	-0.05	-8.06
Absolute Risk	109.03%	42.89%	-66.14%	-60.66	17.68%	23.58%	5.90%	33.35

Green is the values that have increased, Red is the values that have decreased.

Source: (1) MALR (2018) (2) ECAPMS (2018) (3) FOSI Model (2018).

An agro-climatic adaptability classification should be established in a form suitable for matching sugarcane varieties with climate and soil resources and sugarcane varieties production cost established according to soil and climatic zone, sufficient to judge whether yields exceed costs. According to the financial and economic analyses, the annual internal rate of return (IRR) became higher than the existing model for Upper Egypt and increased by 154.498. Absolute risk of optimum cultivation reduced by -139.122% in Upper Egypt (Table 10.3). Table 10.4 shows sugarcane factories efficiency for optimum cultivation based on suitable soil type, laser land leveling in old and new land of Nile valley in Upper Egypt and water and it is comparison with existing condition of Upper Egypt, the results show that the all factories are working by 100% efficiency.

The proposed model provided low greenhouse gases emission than the existing model for all agriculture operations. Pollutant causes destruction of ecosystem, damage to structures and people's health. The social cost of each ton emission of greenhouse gases and air pollutants was accounted on data of optimal use of energy in Upper Egypt in Table 10.5. Finally, laser land leveling should make by the farmers because it is the best solution for the Egyptian question, it is low coast (261.904 Egyptian pounds per hectare in south Egypt), high benefits (5484.612 E.P./hectare) and save water consumption by 46.452%.

TABLE 10.4 Optimum New Efficiency of Sugarcane Factories, Honey Factories, Seed for Planting and Juice Shops in Upper Egypt Corresponding to (FOSI) From Mean (2011/2012–2015/2016)*

Sugarcane industry in Upper Egypt	Abou-korcas sugarcane factory			Gerga sugarcane factory			Qena to Guhag to D	
	Mean	FOSI	Change %	Mean	FOSI	Change %		
Sugarcane factory area	8747	15559.6	77.885	14060	14317	1.828	4258	154
Sugarcane productions (tons)	339412	700033	106.249	552612	1000170	80.989	178964	5838
Sugarcane per fedan for factory (tc	38.803	44.990	15.945	39.304	53.845	36.997	42.03	37.91
Sugar percent %	10.150	12.625	24.383	9.880	12.625	27.778	10059	
Sugar productions (tons)	34451	88379	156.539	54600	126271	131.265		
Efficiency of factory %	48.49%	100.00%	106.249	55.26%	100.02%	80.989		

Sugarcane industry in Upper Egypt	El-menia			Suhag		
	Mean	FOSI	Change %	Mean	FOSI	Change %
Sugarcane factory area	8747	15559.6	100.00%	9956	14471	100.04%
Honey factories	6123	5280.9	100.00%	0	0	0.000
Seed	725	625.3	100.00%	262	191	100.00%
Juce	20503.1	17683.4	100.00%	3932	2870	-26.94%
Total	36098.1	39149.3	100.00%	14150	17533	100.00%

Sugarcane industry in Upper Egypt	Nagy-hamady sugarcane factory			Deshna sugarcane factory			Quss sugarcane factory			Qena to
	Mean	FOSI	Change %	Mean	FOSI	Change %	Mean	FOSI	Change %	
Sugarcane factory area	36750	32888	-10.510	20730	19346	-6.678	42730	30970	-27.522	4258
Sugarcane productions (tons)	1430488	1700000	18.841	757429	1000000	32.026	1519286	1600875	5.370	178964
Sugarcane per fedan for factory (tc	38.925	51.691	32.797	36.538	51.691	41.473	35.555	51.691	45.382	42.03
Sugar percent %	10.035	12.625	25.810	10.350	12.625	21.977	10.445	12.625	20.871	
Sugar productions (tons)	143550	214625	49.513	78396	126250	61.041	158690	202111	27.362	
Efficiency of factory %	84.15%	100.00%	18.841	75.74%	100.00%	32.026	94.96%	100.05%	5.370	

Sugarcane industry in Upper Egypt	Qena		
	Mean	FOSI	Change %
Sugarcane factory area	89258	83203	100.00%
Honey factories	5273	3970.72	100.00%
Seed	4184	3151	100.00%
Juce	19373	14589	100.00%
Total	118088	104913	100.00%

TABLE 10.4 *(Continued)*

Sugarcane industry in Upper Egypt	Armant sugarcane factory			Edfou sugarcane factory			Luxor to	Luxor to
	Mean	FOSI	Change %	Mean	FOSI	Change %		
Sugarcane factory area	36000	32340	-10.167	37900	27365	-27.797	9342	15065
Sugarcane productions (tons)	1301056	1300205	-0.065	1229093	1100189	-10.488	296895	
Sugarcane per fedan for factory (to	36.140	40.204	11.244	32.430	40.204	23.973	31.781	
Sugar percent %	10.601	11.200	5.655	10.627	11.150	4.918	11.150	
Sugar productions (tons)	137918	145623	5.586	130620	122671	-6.086	33104	
Efficiency of factory %	100.08%	100.02%	-0.065	111.74%	100.02%	-10.488	26.99%	

Sugarcane industry in Upper Egypt	Luxor		
	Mean	FOSI	Change %
Sugarcane factory area	60398	56747	100.00%
Honey factories	0	0	100.00%
Seed	2038	1832	100.00%
Juce	4986	4482	100.00%
Total	67422.11	63061	100.00%

Sugarcane industry in Upper Egypt	Edfou sugarcane factory- Aswan			Koom-embou sugarcane fa		
	Mean	FOSI	Change %	Mean	FOSI	Change %
Sugarcane factory area	37850	29810	-21.242	54700	25225	-53.885
Sugarcane productions (tons)	1288065	1300146	0.938	1942971	1100174	-43.377
Sugarcane per fedan for factory (to	34.031	43.614	28.162	35.521	43.614	22.787
Sugar percent %	11.147	11.147	0.000	10.588	10.588	0.000
Sugar productions (tons)	143583	144930	0.938	205718	116484	-43.377
Efficiency of factory %	117.10%	100.01%	-14.591	107.94%	100.02%	-7.344

Sugarcane industry in Upper Egypt	Aswan		
	Mean	FOSI	Change %
Sugarcane factory area	60398	45947	100.00%
Honey factories	0	0	100.00%
Seed	2038	1590	100.00%
Juce	4986	3890	100.00%
Total	67422.11	51428	100.00%

Green is the values that have increased, Red is the values that have decreased.

Source: (1) MALR (2018) (2) ECAPMS (2018) (3) FOSI Model (2018).

It was noted that the sugarcane cultivation season would lose its acreage by 13.773%. As a result of optimal sugar cropping patterns, farm income would increase by 34.275%, water use decrease by 46.452%, CO_2 emission reduce by 31.123%, and energy reduce by 29.901%. Overall, as a result of an optimal sugar-cropping pattern, Egyptian sugar productions would increase by 21.712%, Egyptian sugar exports would decrease by

$94.741 million US. The aim of FOSI model is to study achieving efficiency and equity in sugar factories and sugar-cropping patterns in Upper Egypt by focusing on the Strategic Water Shortage Preparedness Plan, introduction methodologies, and specific action to fight drought within the general water-planning framework.

TABLE 10.5   Changes Crop Emission in Cultivation in Sugarcane Varieties in Upper Egypt Zones Flow Value From Mean (2011/2012–2015/2016) to (FOSI)*

	Sugar cane varieties in El-menia				Sugar cane varieties in Suhag				Sugar cane varieties in Qena			
	Mean	FOSI	Change	%	Mean	FOSI	Change	%	Mean	FOSI	Change	%
NOx	0.012	0.010	-0.002	-17.783	0.004	0.004	0.000	-9.507	0.041	0.028	-0.013	-32.326
SO2	0.057	0.047	-0.010	-17.783	0.021	0.019	-0.002	-9.507	0.197	0.133	-0.064	-32.326
CO2	57.663	47.409	-10.25	-17.783	21.011	19.014	-2.00	-9.507	197.591	133.718	-63.87	-32.326
SO3	nugatory	nugatory	nugatory	nugatory	nugatory	nugatory	nugatory	nugatory	nugatory	nugatory	nugatory	nugatory
CO	0.018	0.015	-0.003	-17.783	0.007	0.006	-0.001	-9.507	0.063	0.042	-0.020	-32.326
CH	nugatory	nugatory	nugatory	nugatory	nugatory	nugatory	nugatory	nugatory	nugatory	nugatory	nugatory	nugatory
SPM	nugatory	nugatory	nugatory	nugatory	nugatory	nugatory	nugatory	nugatory	nugatory	nugatory	nugatory	nugatory

	Sugar cane varieties in Luxor				Sugar cane varieties in Aswan			
	Mean	FOSI	Change	%	Mean	FOSI	Change	%
NOx	0.020	0.016	-0.003	-17.080	0.026	0.013	-0.013	-49.406
SO2	0.094	0.078	-0.016	-17.080	0.126	0.064	-0.062	-49.406
CO2	94.839	78.640	-16.20	-17.080	126.760	64.132	-62.63	-49.406
SO3	nugatory	nugatory	nugatory	nugatory	nugatory	nugatory	nugatory	nugatory
CO	0.030	0.025	-0.005	-17.080	0.040	0.020	-0.020	-49.406
CH	nugatory	nugatory	nugatory	nugatory	nugatory	nugatory	nugatory	nugatory
SPM	nugatory	nugatory	nugatory	nugatory	nugatory	nugatory	nugatory	nugatory

*Nitrous oxide (N_2O), Sulfur dioxide (SO_2), Carbon dioxide (CO_2), Sulfur trioxide (SO_3), Carbon monoxide (CO), Methane (CH_4), and Suspended particulate matter (SPM).

** Green is the values that have increased, Red is the values that have decreased.

Data source: (1) Ozkan et al. (2004); Erdal et al. (2007); Esengun et al. (2007); Bojacá and Schrevens (2010); Mobtaker et al. (2010); Mohammadi and Omid (2010); Rafiee et al. (2010) and Samavatean et al. (2011); (2) MALR (2017); (3) ECAPMS (2018); (4) FOSI Model (2017).

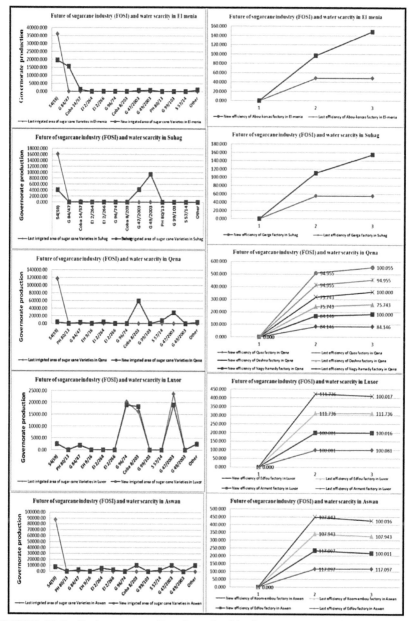

FIGURE 10.6 Optimum area allocation of cultivation sugarcane varieties in Upper Egypt zones and new efficiency of sugarcane factories in Upper Egypt corresponding to (FOSI) from mean (2011/2012–2015/2016), MALR (2018), ECAPMS (2018), and FOSI Model (2018).

10.13   CONCLUSION

A fast-growing population, dwindling access to already scarce resources, and a struggling economic system threaten Egypt's ability to achieve long-run food security. In short, Egypt's capacity to feed itself has become an assignment and a first-rate societal burden considering growing poverty and malnutrition levels together with stagnant food production. In 2015, 27.8% of the population in Egypt changed into residing under the country-wide poverty line, up significantly from 16.7% in 1999/2000. Additionally, 28.8% of youngsters (0–17 years old) were residing in extreme financial poverty in 2013, up from 21% in 1999/2000. Consequently, malnutrition-primarily based growth stunting affected approximately 21.1% of kids beneath 5 and the incidence of anemia become 27% among children in 2014.

Part of the mission for food security is based totally on escalating demand as Egypt's populace grew to 95 million people beings in 2017, up from 73 million people beings in 2006. Given Egypt's excessive fertility price, the populace growth rate increased from 2.04% annually at some point of the duration 1996–2006 to 2.56% annually at some stage in the period 2006–2017. Moreover, 97% of the population inhabits only 4% of Egypt's total region in the Nile Valley and the Delta. Assuming no change (constant fertility and constant mortality), Egypt's population is envisioned to growth to 123 million through 2030 and further to 174 million via 2050. This growing population will aggravate food insecurity no longer handiest with the aid of growing the call for food but also through exacerbating the burden and access to agricultural resources considering development pressure (housing and urbanization) has historically resulted inside the lack of scarce fertile lands, and the degradation of restricted water resources. With 55.5 billion cubic meters (BCM) of water flow a year, the Nile presents Egypt with about 96% of its total renewable water resources. The ultimate 4% comes from groundwater aquifers and only small proportion outcomes from the scarce rainfall received. The agricultural sector consumes most of the country's available water, accounting for 79% of the nation's general water withdrawals because of cropping intensity and inefficient water use. Therefore, water scarcity is a vast danger to food production and food security in Egypt for the country is extensively encountering over-exploitation of the river basin and the water use structure within the MENA region (Middle East and Northern Africa) is predicted to shift far from agricultural uses towards non-agricultural sectors in the next decades.

By 2025, Egypt is anticipated to suffer from physical water scarcity, and grains productiveness is predicted to decrease by 11% as compared to 1995 because of irrigation water shortage. Moreover, latest water infrastructure changes within the Nile Basin, along with the construction of dams and different water projects, could bring about a decrease in Egypt's proportion of the Nile's flow. Furthermore, water pollution is worsening the water crisis and undermining agricultural production as it decreases the amount of suitable water to use. In 2016, approximately 18.9 billion m3 of wastewater (untreated municipal wastewater, business wastewater, and agricultural runoff) became discharged directly into the Nile or into agricultural drains to be recycled lower back to the Nile. This progressively deteriorates soil fertility. For instance, soil salinization is found on approximately 35% of the rural lands in Egypt, limiting the cultivated crops in the salt-affected lands inclusive of the Lower Delta to salt-resistant crops along such as rice.

Arable and fertile lands in Egypt are scarce and farming is confined to less than 4% of the full land vicinity. In 2016, overall agricultural land vicinity changed into about 9.1 million feddans (One feddan equals 1.038 acres or 0.42 hectares) (3.8 million hectares) accounting for 3.82% of Egypt's area. Additionally, land fragmentation into small units poses another assignment to agriculture in Egypt considering smaller farms do now not generate sufficient earnings and are not without problems aggregated into greater commercially possible plots that are higher suited to fashionable manufacturing and supply chains. A growing populace, particularly in rural areas, and the usage of land especially as a de-facto social safety net in Upper Egypt have caused maximum land plots to decrease to tiny subsistence farms. Recently, encroachment on agricultural lands has come to be one of the most hard and pressing problems in Egypt. A report via the Egyptian Ministry of Agriculture and Land Reclamation (MALR) estimated that Egypt had misplaced 326,000 feddans (138,000 hectares) of arable lands among 1983 and 2018 because of development stress and urbanization. Between 2011 and 2018 alone, the country reported 85,000 feddans transitioning out of agriculture.

Water shortages and a shrinking base of arable lands undermines food production goals in Egypt and could sharply growth the reliance on food imports if the impact of climate change is also considered. Most current research predicts a negative effect of weather change on primary crops yield inclusive of wheat, rice, maize, soybeans, and barley. In 2017, Egypt imported 65% of its total wheat consumption, in comparison to 43% in 2013, 53% of maize overall consumption, in comparison to 43% in 2013, and 6% of rice consumption in comparison to no imports in 2013. Reliance on food imports makes the country at risk of commodity rate volatility and calls for a healthful economy capable of earning sufficient foreign exchange to cover the charges of important food imports.

This chapter provides a brief overview of background of future of sugarcane industry in Upper Egypt as a case study of agribusiness and Future of sugar industry (FOSI) and water scarcity in Upper Egypt as a case study of agribusiness.

KEYWORDS

- **clean growth mechanism**
- **emissions**
- **emissions transfer schemes**
- **future of sugar industry (FOSI)**
- **geopolitics**
- **regressive tax**

REFERENCES

Allan, J. A., (2001). Virtual water, economically invisible and politically silent, a way to solve strategic water problems. *Int. Water Irrigate, 21*(4), 39–41.

Al-Suefi, (1974). ABD AL-RAHMAN AL-SUFI (AZOPHI).

Alvaro, C., Katrin, R., Richard, B., Pete, F., Andy, W., & Richard, S. J. T., (2011). Climate change impacts on global agriculture. *Climatic Change, 120*, 357–374.

Annan, K., (2009). *The Anatomy of a Silent Crisis,* Global Humanitarian Forum. Chair, Steering Group, Human Impact Report: Climate Change.

Anonymous, (2008). *Ministry of Agriculture and Land Reclamation Council of Sugar Crops* (p. 28). Sugar crops and sugar production in Egypt, Cairo.

Bluemling, B., & Yang, H., Pahl-Wostl, C., (2007). Making water productivity operational—A concept of agricultural water productivity exemplified at a wheat-maize cropping pattern in the North China plain. *Agric. Water Manage, 91*, 11–23.

Bojacá, C. R., & Schrevens, E., (2010). Energy assessment of peri-urban horticulture and its uncertainty: Case study for Bogota Colombia. *Energy, 35*, 2109–2118.

CA (2007). Comprehensive Assessment of Water Management in Agriculture, Water for Food, Water for Life: Comprehensive Assessment of Water Management in Agriculture, International Water Management Institute, London, Earth scan, and Colombo.

CAN (2007). "National security and the threat of climate change," Center for Naval Analyses and the Institute for Public Research, Alexandria, Virginia.

Charmes, J., (2000). *African Women in Food Processing: A Major, But Still Underestimated Sector of Their Contribution to the National Economy.* Paper prepared for the International Development Research Centre (IDRC), Ottawa, IDRC.

David, Molden (2007). Comprehensive Assessment of Water Management in Agriculture, International Water Management Institute, Page 16, www.earthscan.co.uk .

Deng, X., Shan, L., Zhang, L., & Turner, N. C., (2006). Improving agricultural water use efficiency in arid and semiarid areas of China. *Agric. Water Manage, 80*, 23–40.

Diao, X., Hazell, P., Resnick, D., & Hurlow, J., (2007). *The Role of Agriculture in Development: Implications for Sub-Saharan Africa.* IFPRI Research Report 153, Washington D.C., IFPRI.

ECAPMS, (2018). *Egyptian Central Agency for Public Mobilization and Statistics, Selected Water Related Statistic.* http://www.msrintranet.capmas.gov.eg (accessed on 18 January 2020).

El-Deeb Mohamed Mahmoud Ibrahim, (1999). *Food Industries in Egypt in the Analysis of the Spatial Organization and Installation and Performance* (p. 586). Al-Anglo – Cairo (In Arabic).

Erdal, G., Esengün, K., Erdal, & H., Gündüz, O., (2007). Energy use and economical analysis of sugar beet production in Tokat province of Turkey. *Energy, 32*, 35–41.

ESCWA, (2009). *Economic and Social Commission for Western Asia* (pp. 10–0242). ESCWA annual report, United Nations, New York, 2010.

EU, (2008). *Climate Change and International Security.* Council of the European Union, Paper from the High Representative and the European Commission to the European Council, 7249/08 Annex.

FAO, (2006). *Food and Agriculture Organization, the State of Food Insecurity in the World 2006.* online: http://www.fao.org/docrep/009/a0750e/a0750e00.htm (accessed on 18 January 2020).

FAO (2007). Food and Agriculture Organization of the United Nations, Challenges of agribusiness and agro-industry development, Available at: www.un.org.ecosoc/files/files/en/2017doc/2017_ecosoc_special_meeting_proposal-3ADI%2B.pdf, 2017 (accessed on 18 January 2020).

Ferré, T., Doassem, J., & Kameni, A. (1999). Dynamics of agricultural product processing activities in Garoua, North Cameroon, Garoua, Cameroon: Institute for Agricultural Research for Development (IRAD)/Pole of Research Applied to Development of Savannas of Central Africa (PRASAC).

FOSI Model (2018). Special model built by Youssef Mohamed Hamada, 2018.

Gunn, L. (2017). National security and the accelerating risk of climate change. Elem Sic Ant, 5, p.30. DOI: http://doi.org/10.1525/elementa.227

Hamada, Y. M. (2018). Special pictures album, Took by Youssef M. Hamada, Egypt, 2018.

Hans, L., & Momtaz El-Said, (2001). Food subsidies in Egypt: Reform options, distribution and welfare. *Food Policy*. doi: 10.1016/S0306-9192(00)00030-0.

Hoekstra, A. Y., & Hung, P. Q., (2005). Globalization of water resources: International virtual water flows in relation to crop trade. *Global Environmental Change, 15*, 45–56.

Jaffee, S., Kopicki, R., Labaste, P., & Christie, I., (2003). *Modernizing Africa's Agro-Food Systems: Analytical Framework and Implications for Operations*. Africa region working paper series no. 44, Washington D.C. The World Bank.

Kamara, A., & Sally, H., (2004). Water management options for food security in South Africa: scenarios, simulations and policy implications. *Development Southern Africa, 21*(2), 365–384.

Lilienfeld, A., & Asmild, M., (2007). Estimation of excess water use in irrigated agriculture: A data envelopment analysis approach. *Agric. Water Manage, 94*(1-3), 73–82.

Mailhol, J. C., Zairi, A., Slatni, A., BenNouma, B., & ElAmani, H., (2004). Analysis of irrigation systems and irrigation strategies for durum wheat in Tunisia. *Agric. Water Manage, 70*(1), 19–37.

MALR, (2018). *Egyptian Ministry of Agricultural and Land Reclamation*. Selected data on costs, prices, and land in production.

Mobtaker, H. G., Keyhani, A., Mohammadi, A., Rafiee, S., & Akram, A., (2010). Sensitivity analysis of energy inputs for barley production in Hamedan province of Iran. *Agriculture, Ecosystem Environ, 137*, 367–372.

Mohammadi, A., & Omid, M., (2010). Econometrical analysis and relation between energy inputs and yield of greenhouse cucumber production in Iran. *Appl. Energy 87*, 191–196.

Montalvo, J., & Ravallion, M., (2010). The pattern of growth and poverty reduction in China. *Journal of Comparative Economics, 38*(1), 2–16.

Muchnik, José (2003). Food, know-how and agro-food innovations in West Africa, Collection of reports from the ALISA project. European Union DG XII, Brussels, 2003.

MWRI, (2018). *Egyptian Ministry of Water Resources and Irrigation*. Selected water resources data 2018.

Ozkan, B., Kurklu, A., & Akcaoz, H., (2004). An input-output energy analysis in greenhouse vegetable production: A case study for Antalya region of Turkey. *Biomass Bioenergy, 26*, 89–95.

Pereira, L. S., (1999). Higher performance through combined improvements in irrigation methods and scheduling: A discussion. *Agric. Water Manage. 40*(2–3), 153–169.

Pereira, L. S., Oweis, T., & Zairi, A., (2002). Irrigation management under water scarcity. *Agric. Water Manage. 57*(3), 175–206.

Pomeroy, R., (2007). *Developing Nations Hit Back on Climate Change*. Reuters, February. http://www.alertnet.org/thenews/newsdesk/L27392364.htm (accessed on 18 January 2020).

Rafiee, S., Mousavi Avval, S. H., & Mohammadi, A., (2010). Modeling and sensitivity analysis of energy inputs for apple production in Iran. *Energy, 35*, 3301–3306.

Rose grant, M. W., Cai, X., & Cline, S. A., (2002). *World Water and Food to 2025: Dealing With Scarcity*. International Food Policy Research Institute, Washington, D.C.

Samavatean, N., Rafiee, S., Mobli, H., & Mohammadi, A., (2011). An analysis of energy use and relation between energy inputs and yield, costs and income of garlic production in Iran. *Renew. Energy 36*, 1808–1813.

Sautier, D., Vermeulen, H., Fok, M., & Biénabe, E., (2006). *Case Studies of Agri-Processing and Contract Agriculture in Africa.* RIMISP, Latin American Center for Rural Development, Santiago: RIMISP.

Seckler, D., Amarasinghe, U., Molden, D., de Silve, R., & Barker, R., (1998). *World Water Demand and Supply, 1990 to 2025: Scenarios and Issues.* Research Report 19, Colombo: International Water Management Institute.

Seleshi Awulachew, A., Lisa-Maria, R., & David, M., (2012). *The Nile Basin: Tapping the Unmet Agricultural Potential of Nile Waters.* International Water Management Institute, Addis Ababa, Ethiopia.

Smith, D., & Vivekananda, J., (2007). *A Climate of Conflict.* International Alert, http://www.reliefweb.int/rw/lib.nsf/db900sid/EMAE-79ST3Q/$file/International%20Alert_Climate%20of%20Conflict_07.pdf?openelement (accessed on 18 January 2020).

Soltan, F. H., & Mahmoud, I. N., (2008). Sugar industry in Egypt, Mini review. *Sugar Tech, 10*(3), 204–209.

UK Mission, (2007). *Energy, Security and Climate.* United Kingdom Mission, UK Concept Paper at UN Security Council OPEN Debate.

UN Agency (2007). "Security council holds first-ever debate on impact of climate change," United Nations Security Council, 5663rd Meeting, April 17. https://unchronicle.un.org/article/un-role-climate-change-action-taking-lead-towards-global-response (accessed on 18 January 2020).

UN (2007). United Nations, The Millennium Development Goals 2007, online: https://www.un.org/millenniumgoals/2015_MDG_Report/pdf/MDG%202015%20rev%20 (July%201).pdf (accessed on 18 January 2020).

UN, Water/FAO, (2007). *World Water Day, 2007.* Coping with water scarcity, Challenge of the twenty-first century, UN-Water.

UNIDO, (2009). *United Nations Industrial Development Organization.* Industrial Development Report 2009, Breaking In and Moving Up, New Industrial Challenges for the Bottom Billion and the Middle-Income Countries, Vienna: UNIDO.

UNIDO, FAO & IFAD, (2008). *The Importance of Agro-Industry for Socio-Economic Development and Poverty Reduction.* New York: UN Commission on Sustainable Development.

Walter Fust, (2009). *The Anatomy of a Silent Crisis, Human Impact Report: Climate Change.* The Anatomy of a Silent Crisis Published by the Global Humanitarian Forum—Geneva © 2009, www.ghf-ge.org (accessed on 18 January 2020).

Watanabe, M., Jinji, N., & Kurihara, M., (2009). Is the development of the agro-processing industry pro-poor? *The Case of Thailand Journal of Asian Economics, 20*(4), 443–455.

WBGU, (2008). *World in Transition Climate Change as a Security Risk.* German Advisory Council on Global Change, http://www.wbgu.de/wbgu_jg2007_engl.html (accessed on 18 January 2020).

Wichelns, D., (2004). The policy relevance of virtual water can be enhanced by considering comparative advantages. *Agric. Water Manage. 66,* 49–63.

Wilkinson, J., & Rocha, R., (2008). *Agro-Industries, Trends, Patterns and Developmental Impacts.* Paper prepared for Global Agro-industries Forum (GAIF), New Delhi.

World Bank, (2007). *World Development Report 2008: Agriculture for Development.* Washington, D.C., World Bank.

World Bank, (2009). *World Development Indicators 2009.* Washington, D.C. The World Bank.

WTO, (2008). *World Trade Organization, 2008, World Trade Report 2008.* Trade in a Globalizing World. WTO, Geneva.

Yumkella, K. K., Patrick M. Kormawa, Torben M. Roepstorff, & Anthony M. Hawkins, (2011). *Agribusiness for Africa's Prosperity, UNIDO ID/440.* Layout by Smith + Bell Design (UK). Printed in Austria.

Zimmer, D., & Renault, D., (2003). Virtual water in food production and global trade: Review of methodological issues and preliminary results. In: Hoekstra, A. Y., & Hung, P. Q., (eds.), *Proceedings of the International Expert Meeting on Virtual Water Trade.* IHE and DELFT, The Netherlands.

Index

Printed and bound by CPI Group (UK) Ltd, Croydon, CR0 4YY

23/10/2024

01777702-0007